BASIC Programming for Electronics

BASIC Programming for Electronics

Mike James

Newnes Technical Books

Newnes Technical Books
is an imprint of the Butterworth Group
which has principal offices in
London, Boston, Durban, Singapore, Sydney, Toronto, Wellington

First published 1984

British Library Cataloguing in Publication Data

James, Mike, *19*—
BASIC programming for electronics.
1. Electronics—Data processing
2. Basic (Computer program language)
I. Title
001.64'24 TK7835

ISBN 0-408-01341-9

Library of Congress Cataloging in Publication Data

James, Mike.
Basic programming for electronics.
Includes index.
1. Electronics—Data processing. 2. Basic (Computer program language) I. Title. II. Title: B.A.S.I.C. programming in electronics.
TK7835.J29 1984 001.64'24 83-19436

ISBN 0-408-01341-9

Filmset by Illustrated Arts, Sutton, Surrey.
Printed in England by Page Bros. Ltd., Norwich, Norfolk.

Preface

Once upon a time an interest in microcomputers relied upon a good practical knowledge of electronics — after all, if you wanted to use a micro you had to build it first. These days, however, the computer user can get away without knowing the first thing about the components that make up his machine and probably won't even possess a soldering iron, let alone be proficient at using it. The ready availability of cheap, off-the shelf micros has resulted in a real split between the software and hardware aspects of computing. And just as lots of programmers don't bother to acquire any skills in electronics, because they do not have to, lots of people who appreciate the inner working of the computer don't bother to learn how to use one. In some ways, if you don't have to *build* it first then there's no real challenge in it!

Both positions are very short-sighted. The real fun in computing, for me at any rate, is in *combining* the software with the hardware aspects. In other words, what interests me is using what I know about electronics, and what is going on inside the computer, to improve my programs and make them produce more sophisticated effects. That level of programming, however, is quite a long way off if you've not yet learnt your first programming language and that is where this book comes in. It sets out to teach you BASIC, the most popular of all computer languages.

There is one other advantage in combining an interest in electronics and computing. There are not many other interests, be they hobbies or vocations or any mixture of the two, where the computer can be as directly useful. In electronics you really do need to do calculations, apply formulae, generate test data and simulate circuits and in all of these important areas a computer can be a real asset — as the example programs included throughout this book aim to demonstrate.

To read this book, you need a computer to sit in front of. So now is the time to overcome your prejudices about buying ready assembled electronic equipment and arm yourself with an off-the-shelf micro. Initially it may not seem as challenging as building it yourself but in the long run I hope you'll find it opens up new and rewarding horizons.

This book originated in a series of magazine articles, entitled Building BASIC, which appeared in *Electronics and Computing Monthly* in 1982–83, and my thanks are due to Dave Raven, its Publisher/Editor, for founding a magazine in which computing and electronics can live together creatively.

M.J.

Contents

1

Getting started

Electronics and computers are so intimately connected that it is often difficult to tell where one subject ends and the other begins. After all, every computer *is* a box full of electronics and there really are no hidden ingredients that the electronics enthusiast could not have met elsewhere.

Computers are one of the most interesting creations of modern electronics and it is only natural for an interest in electronics to extend into computing. It often works the other way round as well. An interest that extends only to the *use* of computers is very limited — you can make things happen but you cannot explain how or why they do so — and the desire to know what goes on inside the box can drive one or two programmers to delve into electronics.

However, in this age of information technology, the need to know how to use a computer is more general than the need to know how it works, and indeed it is almost unthinkable that anyone who understands how a computer works would avoid learning how to use one.

It might seem that the electronics enthusiast would have a head start when it comes to using a computer because he is already familiar with all its constituent parts. This is, unfortunately, not the case. The trouble is that a working computer looks nothing like electronics. In fact the manufacturers go to a great deal of trouble to make sure that the innards are never seen by the user. The use of a computer actually involves the developing of an entirely new skill — that of programming.

However, there is a definite advantage in combining electronics and programming. If you do know something about electronics, both analog and digital, and then go on to master programming, you will be in a much better position to use computers creatively than if you are 'just a programmer'. After acquiring both skills, you will know both how to use a computer and how and why it performs in the way it does.

The purpose of this book is to take what you know about electronics and add to it the programming skills necessary to get the best (or anything at all) from a computer. On the way to this objective many of the uses of the computer in electronics will be described and you may even learn a little more electronics! The freedom that a computer gives for carrying out complicated calculations without any effort and with amazing accuracy is very important to the future of electronics. On the grandest scale, computers can be used to design integrated circuits and even whole electronic assemblies, from PCBs and components to layouts and finished boards. Ultimately, no doubt, they will even design the robots that will put all the rest together. At a more humble level, computers can carry out routine calculations such as selecting a current-limiting resistor for a Zener diode. In some ways these humble applications are even more important. By using a computer to work things out, not only can you ignore arithmetic but to a certain extent you can also ignore much of the unnecessary symbol shifting that puts people off algebra. It could be said that, just as the electronic calculator banished arithmetic, the computer will make the ideas that lie behind difficult algebra more accessible. Writing programs and working things out using a computer can be a very good way of gaining confidence in electronics.

If you haven't got access to a computer then the best advice is to buy one! And even if you do have access to a computer then it is still a good idea to consider buying one of your own — the investment will repay itself over and over again. Even if you're not entirely sure about all this software business, now is the time to take the plunge, because programming is an *activity* and if you read on in this book without a computer at your side then you are making learning more difficult than it need be! All of the discussion and program examples used will apply to one of the cheapest machines available — the Sinclair ZX81. Although this isn't a 'program the ZX81' book you could do worse than go out and buy such a machine (currently retailing for less than £50 new, and there are even cheaper second-hand ones readily available). After all, a traditional course in BASIC would cost more than the ZX81!

If you've already taken the plunge and are worried about all this talk of the ZX81 because it's not the one you've bought then be reassured that this book is also for you. Equally, if you want to buy a micro with more features, such as colour and sound, then do not feel restricted in your choice. Other suitable machines (although more expensive) include the BBC Micro, ZX Spectrum, Dragon, Oric, any of the older stalwarts such as the Pet, Apple, Tandy TRS-80 Models I, II and III, and almost any other CP/M machine with a version of Microsoft BASIC.

The choice of micro is wide because BASIC is now a fairly standard language and is implemented in very similar versions on almost every micro you could mention. This means that the form of BASIC used in this book will be similar to that found on most popular microcomputers. Any oddities of BASIC on the popular machines will be pointed out in the course of the book, as we go along.

The beginning

So these days most microcomputers run BASIC in one form or another. BASIC is however just one of a number of high level languages that are available. To know what is meant by 'high level language' we first have to realise that today's microprocessor is, in computing terms a very simple device. The actions that it is capable of carrying out are small and insignificant when matched to the sort of things we usually want to do. For example the (to us) very simple sum

1.5 × 2.5

will require the typical microprocessor to carry out literally hundreds of its simple operations. The point is that the micro is only capable of doing in one go very micro things. Its strength comes not from its incredible power or cleverness but from its ability to work very, very quickly. A single micro action can take as little as one microsecond so that means you can get up to one million micro actions packed into one second. Our simple sum might mean that the micro has to do a thousand operations but it would still only take about one thousandth of a second to complete. In other words the micro is very stupid but very fast. This is no disadvantage, but which would you rather do: write down one thousand commands or 1.5 × 2.5?

This is where a high level language comes in. A high level language is a way of specifying (using English-like terms and arithmetical symbols) what you want the micro to do in such a way that the micro itself can translate your wishes into a large collection of its own tiny actions. The resulting list of its own tiny actions is usually referred to as 'machine code', which is said to be a low level language. So we write our instructions to the machine in a high level language and use the machine to translate to the low level language that it understands directly. (More of the translation process later when we consider compilers and interpreters in Chapter 9.)

BASIC is a high level language invented long before the micro was even thought of. It was originally intended as a simple teaching language. BASIC stands for Beginners All-purpose Symbolic Instruction Code. The idea was that you'd learn to program in BASIC and then move on to some more sophisticated language such as FORTRAN or ALGOL. (At that time these two were generally considered to be sophisticated.) The trouble was that BASIC was such an easy language to use that people just didn't want to move on to another language. The result was that lots of extras were added to the initial form of BASIC to give a better language. The way that these extras were added varied from place to place, and so these days we are left with a core of BASIC that is the same everywhere and extras that may be different. Thus the particular type, or dialect, of BASIC that you use depends on which computer you use but we at least can all agree on a few central statements.

Along with the micro came new versions of BASIC, the best known and most used of which was written by a software firm in the USA — Microsoft. Microsoft BASIC is used by most of the well-known micros — the Pet, Oric, TRS-80, Video Genie, Apple, etc. The only well-known machines that do not use Microsoft BASIC are the Sinclair ZX81, the ZX Spectrum and the BBC Micro, but even these use very similar BASICs. It is because Microsoft has become so popular that this book can teach BASIC in a way that is relevant to so many different machines.

Getting started

Although the BASIC which most micros run is very much the same, the ways in which they run it differ greatly. From here on we will assume that your machine not only runs BASIC but that you know how to get it started. On some machines this is very simple. For example, for the ZX81 and ZX Spectrum all you have to do is plug them in. On other machines this might involve you in loading a machine code program. For example, some Apples have to have their BASIC loaded from tape and the Superbrain needs both the CP/M operating system and BASIC to be loaded from disk.

Whatever has to be done you'll find the details in your machine's manual, often, as here, under the section headed 'Getting started'. While on the subject of manuals it is also worth having your machine's BASIC manual close at hand while reading the rest of this book because, in all points of detail, it is likely to be (but not always!) more authoritative than this book. When you read about a topic in the forthcoming chapters it is a good idea to read the relevant section in the manual both to check that your version of BASIC does work as explained and, more importantly, so that you become familiar with reading the manual. (Reading computer manuals is a skill rather like speaking backwards; few ever persevere long enough to acquire it!)

Entering commands

Another little difference between micros is the way you let the machine know that you have finished typing in an instruction along with the necessary corrections, etc. On many machines all you do is press the key marked RETURN or CARRIAGE RETURN. Of course there are many ways to label a key and I have seen ENTER, ACCEPT and NEWLINE (this last version on the ZX81, the ZX Spectrum and the Video Genie) all used as ways of marking the key that you have to press to say, 'I have finished typing and correcting the line'. So it's up to you to discover which key is the one you have on your machine and from now on we will assume that you will press it at the end of every line you type. What the machine does after you press the Return key depends of course on the line you typed. But if nothing seems to happen remember to 'Return'.

On keyboards and errors

As well as a special Return key most computers have a range of keys that would put a typewriter to shame. The basic layout of all keyboards is the same but the placing of the extra keys that computers use has never been standardised and as a result they are all different. Even if you are a skilled typist you are likely to experience some trouble finding your way around a new keyboard and if you are a 'one-fingered' typist you will find it even harder. It is important that you do not despair at this early stage because learning to use a keyboard is one of the most useful skills that you can acquire and you will be surprised at how quickly you learn where everything is.

A mistake that many beginners make is to confuse any difficulties in using the keyboard with difficulties in using the computer. If you mis-type BASIC commands then at best they will do nothing at all and at worst they will do something drastic that you certainly did not intend. All beginners, and even experts when they use an unfamiliar computer, go through the 'fumble-finger' stage and it is important not to confuse typing errors with programming errors — they are as different as being able to hold a pen and making spelling mistakes. Be assured, though, that you cannot possibly damage your computer with either type of error.

When you try typing in the examples from this book then any errors that you find should be entirely due to mistyping — so before you assume anything worse check that you have entered the program correctly. If you have checked and the program still doesn't work then the most likely solution is that your version of BASIC is somehow different from the version used here. (This is very unlikely but it is just possible that you are using a machine that is not as well known as the ones mentioned earlier.) In this case check your BASIC manual and try to find out what is different.

Although there are many examples given throughout this book there is no way that you can learn BASIC without practice and the only suitable practice is to write your own programs. You will find various suggestions for programs that you can write or improve on in the self-test questions at the end of most of the chapters. It is important that you do take these problems seriously! None of them are difficult (indeed there is nothing difficult about BASIC) and they will make sure that you are following the ideas presented.

Now take out your computer and learn BASIC . . .

2

Numbers and arithmetic

Often the first thing that a computer is thought of doing is arithmetic. In fact the average computer spends a very small proportion of its working life doing sums. What it does do will become clearer as we proceed. However, it is an important first step to discover how to handle numbers accurately. This will in turn lead us on to other topics in BASIC.

Operations and order

In BASIC the usual arithmetic operations are allowed and the only change from the conventional symbols is the use of a '*' sign to mean multiply. The reason for this is that the more normal '×' sign can be confused with the letter 'x' and one thing high level computer languages must be is unambiguous. Also the only sign for division that is allowed is '/'; after all, you only need one!

Having settled on the four operators, it is apparent that, for example, 4 + 3 or 5 + 6/3*4.55 or 6.7/5.8 + 5.6 or −5/6*33.4 + 6.7 are all valid pieces of BASIC arithmetic. But some ambiguity remains. Let's just try working out a simple sum by hand:

```
4 + 3*3
```

Should the answer be 21, which is four plus three (i.e. seven) times three (i.e. twenty-one) or should it be 13, which is three times three (i.e. nine) plus four? If we work out all our sums strictly from left to right the first answer would be correct. Alternatively, if we do all our multiplications first, no matter where they come in the order in which the sum is written down, then 13 is the answer.

Instead of arguing about the correct answer it is important to realise that it is up to everyone to *agree* on how to work things out. Arithmetic was not handed down from upon high — it is an invention of man and *we* decide how it should be done. It happens to be the case that we have agreed to work out multiplications first and hence the second answer — 13 — is correct. It may seem more complicated than necessary but we (that is mathematicians) have decided that arithmetic is not done strictly from left to right but rather that * and / should be done before + and −. Another way of saying this is that * and / have a 'higher priority' than + and − and that we do operations of the highest priority *first*. When we have to choose between signs that have equal priority then we do the operations from left to right.

There is only one more detail that need concern us at this stage, namely that there are two uses of the minus sign. The usual minus occurs between two numbers, for example 5 − 4, and is therefore referred to a 'binary operator'. The other use of the minus sign is in front of *one* number, e.g. −5. This is usually referred to as a 'unary minus' and simply means 'replace the number by its negative'. These two uses of the minus sign are confusing and it might be an idea if one of them was indicated by another sign, but this would make otherwise normal arithmetic look very peculiar. As the two meanings of minus can always be distinguished very easily in practice we can carry on using one sign to mean two things. However, we must be careful to notice that a unary minus has a higher priority than *,/,+ or a binary minus. For example, try 5*−4. (Answer −20.)

If this idea of giving the arithmetic operators different priorities seems difficult to you then you can always use brackets to make the order in which you want things worked out absolutely clear. Any arithmetic within a pair of brackets will always be worked out first. For example, if you want the addition in 3 + 2*2 to be carried out first then put brackets around it, i.e. (3 + 2)*2 is 10. Brackets can be unnecessary, as in 3 + (2*2), but they are never harmful. So, if in doubt use brackets.

You may feel that this problem of the order that arithmetic is done is something that will never give you any trouble. I hope this is the case but it is worth pointing out that one of the most common errors found in BASIC is the incorrect transfer of a formula from paper to computer. For example:

$$\frac{1}{2\times 3}$$

is often incorrectly translated to BASIC as 1/2*3. The reason why this is incorrect should be obvious: the division is done first followed by the multiplication, giving (1/2)*3, but you have to admit that it looks more like the way that it is written on paper than the correct BASIC expression which is 1/2/3 (i.e. one divided by two, and then divided by three). If you want to use something that looks a little more like the original equation, then you have no choice but to use brackets, as in 1/(2*3).

In the sort of arithmetic that is necessary for electronic calculations we need quite a bit more maths than just the simple four arithmetic operators. But don't worry — the computer does most of the work and leaves you free to understand what's going on. The extra 'advanced stuff' will be introduced as we go along.

A big pocket calculator

If you have your computer ready then type in:

```
PRINT 2*2
```

(Remember to press the Return key when you have typed the line.) If your computer is working the answer 4 will appear on the screen. What we have just discovered is that your relatively cheap computer can be used as a very expensive calculator. You can type in any arithmetic you please and the machine will oblige with the answer. Try

```
PRINT 4 + 4*3
```

and you should get the answer 16, showing that your machine obeys the priority rules we have just discussed.

You have also discovered that your computer can take a command from you in BASIC (PRINT is a BASIC command) and carry it out at once. This is the so-called 'immediate mode' of operation. Now let's try

```
10 PRINT 4 + 4*3
```

This time nothing happens when you press Return. The computer just sits there waiting for you to type another line. This is because we have started the line with a number — the 'line number' — and this causes the command to be stored in the machine's memory but not to be acted upon, i.e. the machine works in a 'deferred mode'. To show that the line was stored in, type

```
LIST
```

and the line should be produced on the screen. To get the computer to carry out the deferred command type

```
RUN
```

and the result 16 should once again appear on the screen.

A BASIC program is nothing more than a list of deferred commands. Each command has a different line number and when the program is executed by typing RUN each command is taken in turn according to the value of its line number — lowest numbers first. So if we now type

```
20 PRINT 2*2
```

followed by

```
LIST
```

we will see both the previous line 10 and our new line 20. If we type RUN the numbers 16 and 4 are printed on the screen. And that is our first BASIC program!

Modes and programs

In the last section we discovered that the computer has two distinct modes — immediate and deferred. This is such an important idea that it is worth dwelling on for a moment. When you first switch your machine on it is continuously examining the keyboard to discover what you have typed. As you type in individual letters they are kept together in store in an attempt to assemble the line of BASIC that you are typing. At this stage the machine is not examining what you are typing, it is merely acting as a sort of note pad. While you are typing in the collection of letters you can delete and make changes as you will, but once you press the Return key things change. The computer now examines the command to discover what you want it to do and you can no longer correct any errors or change your mind.

The first thing that the computer looks for is the presence or absence of a line number. If the command lacks a line number then the machine will carry it out immediately. However, if the command begins with a line number then it is not processed any further. It is simply added to any previous commands with line numbers already stored in memory. In this way a list of deferred commands can be built up for later use. Of course this list of deferred commands would be quite useless unless there was some way of telling the computer to obey them. As we have already discovered, the immediate command RUN will make the computer start obeying the list of instructions in order of increasing line number. We could say that while the computer was obeying this list of instructions it was in yet another mode, because it is no longer examining the keyboard to see what you are typing — it is getting on with its own set of internal jobs.

Editing

A program can be built up by typing in deferred BASIC commands. Each command can be edited before pressing the Return key by using the Backspace or Delete key. However, after the Return key is pressed the computer stores what you have typed. This is fine as long as you always enter perfect commands! The most straightforward way of altering a statement once it has been entered is to type the whole thing in again. If you type in a line with a line number that already exists then your new line will replace it. In other words, typing 10 PRINT 3*3 will add the new line to the program even if it means overwriting an existing line 10.

If you type a line with a line number in between two numbers that already exist then your new line will be inserted into your program at the correct place. So if you have a program that consists of lines 10 and 20 and you want to insert an extra command between them simply use line number 15. If you want to delete a line simply type in its line number and nothing else. That is, typing 10 'Return' will remove line 10 if it exists in the program.

If you want to delete your entire program then you can type the immediate command NEW. But take great care — on most machines there is no way of undoing the effect of NEW. After each of our examples, and before the next one is typed in, you must remember to delete the old program by typing NEW. Otherwise all the examples will get mixed up!

There are more sophisticated forms of program editing but these vary from machine to machine. It is however well worth finding out about your computer's editing commands. Some popular micros have facilities that make editing very straightforward while others are less easy to use. Even so, once you get used to using them they do make entering long programs much easier.

BASIC commands

The first stage in learning a computer (or any other) language is to acquire a few words. This is what we'll do with BASIC; introduce a few simple commands and then go on to write a small program. Only the simplest form of each command will be considered at this early stage so this means that we will have to revisit each command as we become more proficient and try to write more difficult programs.

The PRINT statement
This is our first BASIC command. The form of the print statement is

```
PRINT print list
```

where 'print list' is simply a list of the things that we want displayed on the screen. The range of things that can be put in a print list is very wide and we will introduce new types as we go along. The most important, however, are 'literal strings' and 'variables'.

A literal string is anything you care to type but with quotes around it. The characters between the quotes, i.e. the string, are printed exactly on the display.

A variable is the name of an area of the computer's memory that is available to you for the storage of data. Some machines allow very long names to be used but only take notice of the first two letters of the name. Others only allow two-letter names, some only one letter and one number (e.g. F3) and the simplest only allow one-letter names. For simplicity we will use only one-letter names, but if your machine can handle more, all the better as it means that you can use meaningful variable names that indicate the nature of what is stored in each variable. Variables are used to store numbers for use in later calculations. Exactly how to store numbers in them is discussed in a later section.

Items in a print statement have to be separated from each other, otherwise we would not know where one began and another ended, and BASIC gives special meaning to how they are separated. If a comma is used then a number of spaces are printed between each item. If a semicolon is used then no spaces are printed.

The time has come for us to try some PRINT statements. Type

```
10 PRINT "HELLO THIS IS A literal STRING"
20 PRINT "ONE","TWO"
30 PRINT "ONE";"TWO"
40 PRINT "A",A
RUN
```

Each PRINT statement illustrates some point. Line 10 simply prints the literal string. Lines 20 and 30 show how commas and semicolons can be used to alter spacing. Line 20 prints something like

```
ONE          TWO
```

but line 30 prints

```
ONETWO
```

Line 40 shows the difference between a literal and a variable; the A between quotes appears on the screen but the second A is not between quotes and is therefore the name of a variable. On a number of machines this line will produce an error message or 'report code'. This is because you have tried to PRINT the contents of an area of memory called A but at no time in the program have you ever stored anything in it. How can you expect the computer to print the contents of a variable that you have never used? However, on most machines line 40 produces

```
A           0
```

Where has the zero come from? 'A' is a variable and PRINT A means print the contents of the computer's memory called A. So what is printed depends on what is in A and BASIC has a convention that all variables are set to zero when RUN is typed.

It is important to realise that this automatic zeroing is something that is far from standard in BASIC. Because of this it is good practice not to assume that a variable has an initial value of zero but explicitly to set it to something. If you follow this rule of always defining the contents of all the variables that you use in a program then you will find that there is more chance of your program working on another version of BASIC.

The INPUT command

INPUT is in some senses the opposite of PRINT. It asks the user to supply information. This information is placed in a variable for use later on by the program. For example

```
10 INPUT A
20 PRINT A
RUN
```

should produce a question mark on the screen. If you type a number, 123 say, and press the Return key then the number that you typed in should be printed on the screen on the next line. INPUT A means 'get a number from the keyboard and save it in the memory area called A', that is in variable A, and we already know what PRINT A does. Notice that INPUT causes the computer to stop carrying out your program and begin examining the keyboard instead. It will carry on examining the keyboard until you have typed in a number and pressed Return. Then it stores the number in the variable named in the INPUT statement and carries on with the rest of your program.

On most computers (but not the ZX81) you can use a more advanced form of the INPUT statement. A full INPUT statement can ask for more than one variable to be input at a time. The general form is

```
INPUT input list
```

The input list is simply a list of variables separated by commas. Now type

```
10 INPUT A,Z
20 PRINT A,Z
RUN
```

When the question mark appears on the screen it is asking for two numbers. You could, for example, type in reply

```
2.3,4.5
```

This would result in

```
2.3          4.5
```

appearing on the next line. Notice that the two numbers were input together by separating them by commas. Some machines will allow you to type in one number at a time and will re-prompt with the question mark for the next. See what your machine does if you re-run the program and enter three or four numbers in one go.

Assignment
The INPUT statement is one way of getting information into variables. The assignment statement is another. Its general form is

```
LET variable = expression
```

where variable can be any valid variable, and expression can be any valid arithmetic expression. Type

```
10 LET A = 4.5
20 PRINT A
RUN
```

The number 4.5 should appear on your display.

Now try

```
10 LET B = 3.14159
20 LET R = 2
30 LET A = B*R*R
40 PRINT "THE AREA OF A CIRCLE OF RADIUS ";R;" IS ";A
RUN
```

If you remember that the area of a circle is pi (3.14159) times the radius squared and that squaring is the same as multiplying a number by itself you should have no trouble with this example. The only new feature is the use of variables within the arithmetic and this is discussed in the next section.

It is worth mentioning that most versions of BASIC (but not ZX BASIC) will allow you to leave out the LET and this can save a lot of typing. Most of the examples in this book will omit LET, so if you are using a ZX machine remember to put it in; if you don't you will get a 'variable not defined' error code message.

Variables

The idea of a variable has so far been introduced as part of the explanation of the BASIC statements INPUT, PRINT and LET. Although not a difficult idea it can prove tricky until you get used to it. A variable is a named area of computer memory that can be used to store data. Any data stored in a variable exists only while the program is running and is most certainly lost if the machine is switched off. You can re-use variables by storing new values in them but any existing contents will be overwritten. For example:

```
10 A = 1
20 PRINT A
30 A = 2
40 PRINT A
```

stores 1 in A, PRINTs it then stores 2 in A and once again PRINTs it. One useful way to think of a variable is as a named 'box' in which you can store a number.

As we saw in the last section, you can store a value in a variable using the assignment statement or the INPUT statement and you can find out what is stored in a variable by using the PRINT statement. However, there is more to a variable than simply storing and retrieving values. You can use a variable in an arithmetic expression just as if it was a number. So in the area of a circle example in the last section line 30 read

```
30 LET A= B*R*R
```

and this worked out the result of multiplying the contents of B and R together. In other words, expressions such as X*Y are worked out using the current contents of the variables X and Y. You can also mix variables and numbers to make expressions such as 3*A/B + 20. In fact you can use a variable anywhere that you would be allowed to use a number.

This all seems very straightforward but you need to keep a clear head to work out the meaning of something like

```
10 A = 1
20 PRINT A
30 A = A + 1
40 PRINT A
```

If you look at line 30 you will see that A appears on both sides of the equals sign. Line 10 stores 1 in A and then line 20 prints it. What do you think the result of line 30 is; i.e. what is printed by line 40? Line 30 simply adds one to the number currently stored in A. At the point when the expression on the right-hand side of the equals sign is worked out, A already contains one and so the result is two. Once this result is obtained it is stored in A, so overwriting the old value, and the value printed by line 40 is clearly two. A variable can appear on the left and right-hand sides of an assignment statement and in this case it is important to realise that the right-hand side is worked out first and then the result is stored in the variable on the left.

Calculating resistance values

As an example of what we can do with the three statements that we now know, let's try to calculate the total resistance of two resistors in series. If we have two resistors of A ohms and B ohms then the total resistance if they are connected in series is simply A + B ohms. Type

```
10 PRINT "WHAT IS THE VALUE (IN OHMS) OF THE FIRST RESISTOR"
20 INPUT A
30 PRINT "WHAT IS THE VALUE OF THE SECOND RESISTOR"
40 INPUT B
50 LET C = A + B
60 PRINT "TOTAL SERIES RESISTANCE =",C
RUN
```

Easy, isn't it?

For resistances in parallel the total resistance is given by a more complex expression A*B/(A + B), but this makes it more worth while writing a program for it! Alter the previous program by typing in the following lines

```
70 LET C = A*B/(A + B)
80 PRINT "TOTAL PARALLEL RESISTANCE =",C
```

You can obviously add to this program for capacitors and inductors but it's more difficult to extend to the case of more than two resistors. You need some more BASIC before that's possible.

Problems with powers

In the area of a circle program given earlier the need arose to calculate the square of a number. The way that this was done was to go back to the definition of a square as X*X. However, there are times when you have to calculate a power that cannot be changed into repeated multiplication. For just this reason most versions of BASIC include a 'raise to the power' symbol. However, there are two common symbols in use: ** (double asterisk) and ↑ (the up arrow). Find out which your version of BASIC uses (we will use ↑ for the rest of this book) and try

```
10 INPUT A
20 INPUT B
30 C = A ↑ B
40 PRINT C
```

(remembering to use the correct symbol in line 30). Use this program to discover what values of A and B your version of BASIC will calculate powers for. It is often not as wide as you might expect because of the limitations of the way powers are worked out. When used with other operations, raising to a power (also called exponentiation) has the highest priority and so will always be carried out first. For example, 3 + 2 ↑ 2 is 3 + (2 ↑ 2) or 7, and 3*2 ↑ 2 is 3*(2 ↑ 2) or 12. You can also raise numbers to the power of an expression, for example 2 ↑ (1/2) will find the square root of 2.

It is important to realise that, although all computer arithmetic is done at lightning speed, powers are generally so difficult to calculate (involving working out logs, multiplication and working out antilogs) that they should be avoided if at all possible. So rather than A ↑ 2 use A*A, rather than A ↑ 3 use A*A*A, and so on. However, all things have to be kept in proportion and A ↑ 50 is certainly to be preferred to writing A fifty times!

Self-test questions

1 Work out the following arithmetic expressions (using a calculator but *not* a computer!)
(a) 2*3 + 11
(b) 3 − 4/8
(c) 6*35/25
(d) 1/2 + 2
(e) 1/(2 + 2)

2 Write a short program to input the length of a side of a square and print its area.

3 Modify your program for question 2 to work out the area of a rectangle.

4 The number of turns needed for an air-cored solenoid of inductance L μH is given by

$$\frac{0.4Ld + 0.16L^2d^2 + 0.0033r^3L}{0.0031r^2}$$

where d is the diameter of the wire in millimetres and r is the radius of the core in millimetres. Write a program that will print out the number of turns needed for any given inductance, wire diameter and core radius. Use your program to investigate the possibility of using an air-cored coil in a tuned circuit needing an inductance of 10μH.

3

Control instructions

By the end of Chapter 2 we knew enough BASIC to be able to write programs. However, although they were very useful programs they were hardly any different from the sort of thing that can be done without very much more trouble on a calculator. In this chapter we come to grips with the very essence of computer programming — instructions that alter the order in which other instructions are carried out.

The flow of control

So far the programs that we have written have been carried out in order of line number, the statement with the lowest line number being executed first. This corresponds to executing the program by reading it line by line from top to bottom. The order in which statements are executed is known as the 'flow of control' because as each statement is executed it is in 'control' of what is happening, and control 'flows' from line to line down the program.

Although this simple flow of control is fundamental to programming it is inadequate for most problems. For example, consider the very simple task of printing out the series of numbers 1, 2, 3, ... Currently the only method that we could use gives something like:

```
10 A = 1
20 PRINT A
30 A = A + 1
40 PRINT A
50 A = A + 1
60 PRINT A
. . .
```

We start with A as 1 and print it in line 20, then add 1 to make A 2 and print it in line 40, and so on. Clearly the program gets very long and very tedious. What we need for this simple problem is a way of repeating the same action: add 1 to A and print A, over and over again. We need a loop.

Looping and GOTO

The GOTO instruction is a way of making your computer carry out an instruction that might not be the next one in the sequence. The form of the GOTO instruction is simply:

GOTO line number

or

GO TO Line number

depending on your computer and which version you like; most computers accept both. (The ZX81 has GOTO written on one of its keys but displays GO TO.) The effect of a GOTO is easy to see in the following little program:

```
10 PRINT "THIS IS LINE 10"
20 GOTO 40
30 PRINT "THIS IS LINE 30"
40 PRINT "THIS IS LINE 40"
50 PRINT "THIS IS LINE 50"
```

The result of running this program is that it is obvious that lines 10, 40 and 50 are executed. This shows that line 30 is skipped as the GOTO 40 is carried out and that after line 40 the flow of control returns to normal — not to the line following 20 but to the line following 40. That is, a computer has no memory of executing a GOTO and unless a GOTO is being obeyed control flows from line to line downward.

How does the GOTO help us to repeat operations? Type:

```
10 A = 1
20 PRINT A
30 A = A + 1
40 GOTO 20
RUN
```

The display of your machine will fill with numbers, one to a line. The GOTO at line 40 transfers control to line 20 and so round the loop we go for evermore! Our only problem is stopping — some machines will stop with an error message when the number gets too big or the screen is full, for others the STOP or RESET key will have to be pressed. Either way is not a good way of stopping!

If you want to see more numbers on the screen try substituting

```
20 PRINT A;
```

If you want to try a different sequence of numbers substitute

```
30 A = A + A
```

or

```
10 A = 2
30 A = A*A
```

This repeating of a very simple action is one of the fundamental methods of programming — its technical name is 'iteration'.

Conditional expressions

Sometimes we want the flow of control to change according to some condition. For example, in the previous counting program a way of stopping neatly is to obey the GOTO only if A is less than 10, say. This would print 1 to 9 and then stop. This can be done using an IF statement. Typing

```
10 A = 1
20 PRINT A
30 A = A + 1
40 IF A<10 THEN GOTO 20
RUN
```

should print 1 to 9 on your display. Line 40 is a typical IF statement. Between the IF and the THEN can be written a wide range of conditions to be satisfied; we will only deal with the simplest for now. If the condition is true then the statement following the THEN is carried out — in this example the GOTO 20. If the condition is not true then the flow of control passes to the next line as usual. In this example the program finishes for lack of more instructions. Thus the general form of the IF instruction is:

IF condition which may be true or false THEN some instruction

Notice that the instruction following the THEN can be any BASIC instruction and not just a GOTO as in our example. In fact the instruction following the THEN can even be another IF in some forms of BASIC.

The condition between the IF and the THEN is usually referred to as a 'logical' or 'conditional' expression. In simple BASIC it takes the form:

arithmetic expression relational operator arithmetic expression

For example, A + 3 = Z is a conditional expression; the arithmetic expressions are A + 3 and Z and the relational operator is =. The standard relational operators in BASIC are

= equals
< less than
> greater than
<= less than or equal to
>= greater than or equal to
<> not equal to

The following examples of IF statements indicate some of the ways in which they can be used:

```
IF A = Z + 3 THEN PRINT "Z IS BIGGER THAN A BY 3"
IF Z<10 THEN Z = A + Z
IF Z<>0 THEN GOTO 50
```

This is a first introduction to the IF statement; we shall return to its more elaborate forms in Chapter 9.

Stopping

Now that we have the IF statement it would be useful if we could write 'IF some condition THEN stop running the program'. What we need is a STOP instruction. The only complication is that some BASICs use END and others use STOP and some use both! So find out which your version of BASIC recognises as meaning 'give up processing'. For the rest of this book STOP will be used.

FOR loops

We have now all the BASIC instructions that we need to write almost any program. However, there are some types of program loops that occur so often that it is useful to have a simple instruction for them. In particular it is often necessary to have the same section of program executed a number of times. For example, to print the fourth power of two — 2*2*2*2 — we could use:

```
10 N = 4
20 A = 2
30 I = 1
```

```
40 A = A*2
50 I = I + 1
60 IF I<=3 THEN GOTO 40
70 PRINT "THE ";N;"TH POWER OF TWO IS ";A
```

In line 10, N is set to the power of two we need, and in line 20 A is initialised to 2. I is a new special variable that we are going to use to count the number of times we go through the loop. Such a counting variable is very often called an index variable. At line 40 we form 2*2 and the first time through the loop I = 1. Line 60 checks to see if we have been through the loop enough times and sends us back if we have been through less than three times. Notice that at line 60 the index I is one *more* than the number of times through the loop. The result is that we do instruction 40 three times giving 2*2 = 4 on the first pass, 4*2 = 8 on the second and 8*2 = 16 on the third.

What we wanted to do was carry out line 40 for I equal to 1, 2, 3. For just this purpose BASIC provides the FOR loop

```
FOR index variable = expression one TO expression two
                 . . .
     any list of BASIC statements
                 . . .
NEXT index variable
```

The two instructions FOR and NEXT together form a FOR loop. On encountering a FOR instruction the computer works out the values of the two arithmetic expressions and sets the index variable to the value of the first expression. The second expression is stored for later use (in an area of memory called the FOR stack). The NEXT instruction simply marks the end of the loop. On encountering a NEXT, the computer increases the index variable by one and, if it is still less than or equal to the value of expression two, control transfers to the instruction following the FOR.

Let's do the power-of-two program again but with a FOR loop:

```
10 N = 4
20 A = 2
30 FOR I = 1 TO 3
40 A = A*2
50 NEXT I
70 PRINT "THE ";N;"TH POWER OF TWO IS ";A
```

which of course produces the same result as the previous example.

Because the expressions in the FOR instruction can be *any* valid arithmetic expression:

```
30 FOR I = 1 TO N - 1
```

can also be used, with the added advantage that we can replace line 10 by

```
10 INPUT N
```

to give a general nth power-of-two program.

By experimentation, find out how large N can be on your computer — and guess before you try.

The index variable within a FOR loop can also be used as a normal BASIC variable subject to one rule: an index variable must *not* be altered during a FOR loop. So

```
10 FOR I = 1 TO 10
20 PRINT I
20 NEXT I
```

is legal, but

```
10 FOR I = 1 TO 10
20 I = 3
30 NEXT I
```

is not.

There are a number of other restrictions which apply to FOR loops. Some depend on which machine you're using but two almost universal ones are:

(a) you may not jump out of an active FOR loop; and
(b) you may not jump into a FOR loop and then encounter a NEXT.

The first restriction means that if during the running of a FOR loop you discover (by way of an IF) that everything is already complete, so that going through the FOR loop would be a waste of time, you still have to go on. (The reason for this is that only when a FOR loop is finished properly is the second expression removed from the FOR stack.) The second restriction is required because if you encounter a NEXT statement as the result of jumping into a FOR loop without executing the FOR, the computer doesn't know what to do 'next'. It is possible to jump into a FOR loop and out again without encountering the NEXT and no errors will occur, but this is usually taken to be very sloppy programming and there is always a better way.

Advanced FOR loops

The form of the FOR loop introduced in the last section serves for most applications but there is an extension to the FOR loop that can sometimes be useful. In a standard FOR loop the value of the index variable is increased by one each time. By using the STEP instruction you can set the amount by which the index variable is increased to anything you want. The general form is:

```
FOR index = expression one TO expression two STEP expression
            three
              ...
      any list of BASIC statements
              ...
NEXT index variable
```

For example, try

```
10 FOR I = 1 TO 10 STEP .1
20 PRINT I
30 NEXT I
```

which is an example of a step size, or increment, of 0.1. That is, each time through the loop the index variable increases by 0.1 until it is just larger than 10. You might like to ask yourself, what is the last value of I to be printed out? There is nothing tricky about using fractional STEP sizes apart from remembering that the FOR loop ends when the index value is just larger than the second expression. However, it is quite possible to specify a negative step size. For example

```
10 FOR I = 10 TO 1 STEP -1
20 PRINT I
30 NEXT I
```

In this case one is subtracted from the index variable each time through the loop. The only other change is that now the loop ends when the index variable is just smaller than the second expression.

Working out when FOR loops end is easy as long as you keep in mind the two rules: (a) for a positive step size the loop ends when the index is just bigger than the second expression; and (b) for a negative step size the loop ends when the index is just smaller than the second expression.

FOR loops can begin to look a little complicated when negative values are used in positions other than the step size but as long as you know that the more negative a number is the smaller it is (e.g. −2 is greater than −3) then the same two rules can be applied. For example

```
10 FOR I = -100 TO -10 STEP +1
20 PRINT I
30 NEXT I
```

will print −100, −99 up to −10 (because −9 is bigger than −10). And replacing line 10 by

```
10 FOR I = −10 TO −100 STEP −1
```

will print −10 down to −100 (because −101 is smaller than −100).

The case of the missing FOR loop
There is one problem associated with using the FOR loop that results from the lack of a standard form of BASIC. Consider (and indeed try) the following short program:

```
10 INPUT S
20 INPUT E
30 FOR I = S TO E
40 PRINT I
50 NEXT I
60 GOTO 10
```

This is a very simple FOR loop with variable start and end values stored in S and E. If you enter values such as 1 for S and 10 for E you will see the range of number that you would expect printed by line 40. However, what happens if you try a value of 1 for S and 1 for E? The answer is, nothing unexpected — if you apply the rules that govern the way that the FOR loop works. The loop is carried out once, resulting in 1 being printed at line 40, because when it reaches the NEXT statement one is added to the index variable making it 2 and thus larger than the value of E.

Now consider the same problem but with a value of 2 for S and 1 for E. This time things are not so clear cut because the index variable starts off with a value that is larger than E (i.e. 2 is greater than 1). Should the FOR loop be carried out once, resulting in line 40 printing the value 2 and then stopping — because the NEXT statement adds one to the index variable, giving 3 which is clearly greater than 1? The answer is in fact unclear and this is reflected in the way some versions of BASIC carry out such a FOR loop once, and other versions not at all. In practice it most often turns out that the FOR loop would be skipped when the initial value is greater than the final value. If you are writing a program where this is the case then using something like

```
10 INPUT S
20 INPUT E
30 IF S>E THEN GOTO 70
40 FOR I = S TO E
50 PRINT I
60 NEXT I
70 STOP
```

will guarantee that your program will work no matter what your BASIC does. It is always worth paying careful attention to FOR loops that have variable initial and final values to make sure that they always do the right thing.

Nesting

As you begin to write more complicated programs you will want to use more than one FOR loop per program. The rules for using multiple FOR loops can seem quite difficult but it's all a matter of nesting. If you have two, or any number of FOR loops in your program, executed so that one ends before the other begins, then you can't go wrong; that is, the pattern

FOR I . . . NEXT I . . . FOR J . . . NEXT J

is always legal. However, if you start a new FOR loop *within* another, then things can go wrong. The rule is that FOR loops cannot overlap — they must be fully nested within one another.

For example, FOR I . . . FOR J . . . NEXT I . . . NEXT J is not acceptable because the first loop ends before the second, and thus the second is not contained in the first. FOR I . . . FOR J . . . NEXT J . . . NEXT I is the correct form. As long as the inner FOR loops are completely contained within the outer FOR loops you can carry on without errors, given that your machine can handle that number of FORs of course. (The ZX81, for example, can handle up to 26.)

Nested FOR loops really only come into their own when we learn something about arrays in the next chapter. If you are impatient to see the effect of a pair of nested FOR loops then try

```
10 FOR I = 1 TO 10
20 FOR J = 1 TO 10
30 PRINT I,J
40 NEXT J
50 NEXT I
```

To understand what is going on it helps to realise that the inner loop — the one that uses J as an index variable — will start at 1 and index its way to 10 for each time through the outer loop — the one that uses I as an index variable.

Using GOTO, IF and FOR

The three BASIC statements introduced in this chapter are very easy to understand. Unfortunately, understanding them is not enough to enable you to write BASIC programs. You have to learn how and when to use them and this is something that only comes with practice. However there are some general points that can speed up this process and hopefully avoid the misuse of the commands.

Use the GOTO statement to form loops where necessary. Do not use it to jump haphazardly around the program — there is usually no need. The IF statement should be used to break out of loops when a condition is satisfied, to skip a section of program or to select between two alternative parts of the program. The FOR loop should be used whenever a section of program has to be executed a given number of times. If a loop has to be carried out until some condition is satisfied then use the GOTO to form a loop and the IF statement to jump out of it — *never* jump out of a FOR loop.

Using the 555 timer

As an example of a *real* program using some of the statements introduced in this chapter what follows will be of use in designing timers using the 555 timer chip. For those not familiar with this useful device, Fig. 3.1 shows a typical 555 timer circuit. A pulse on the trigger line (i.e. closing the switch) results in the output going high for a time given by:

$$T = 1.1 \times R \times C \text{ ms}$$

with R in kilohms and C in microfarads.

Designing a timer is obviously a simple job. All we have to do is find values for R and C that give us the required time T. However, the following limitations of the circuit have to be allowed for:

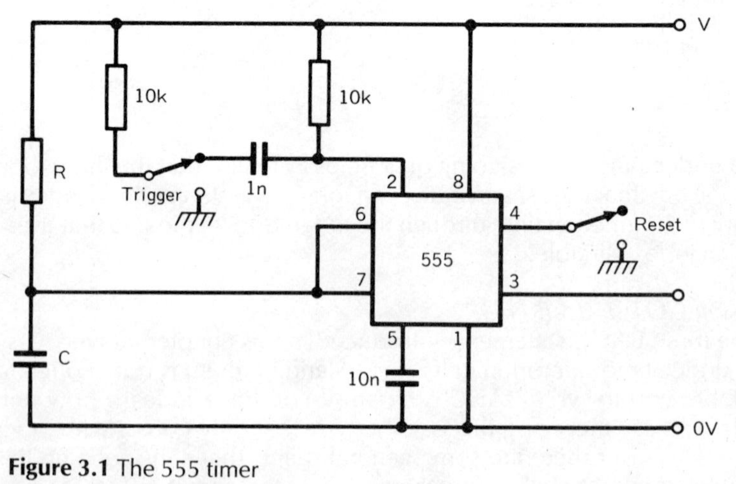

Figure 3.1 The 555 timer

(a) R must be greater than 5 kilohms; and
(b) R must be less than 3.3 × V megohms (where V is supply voltage).

Also it would be useful if the values of R and C were easy to obtain. To this end, it would be a good idea if we could examine the range of values that would give us our required T.

With these requirements in mind, here is our program.

```
10 PRINT "WHAT TIMING PERIOD - IN MILLISECONDS DO YOU REQUIRE ";
20 INPUT T
30 PRINT "WHAT IS YOUR SUPPLY VOLTAGE - IN VOLTS";
40 INPUT V
50 A=T/(1.1*5)
60 B=3.3*V*1000
70 C=T/(1.1*B)
80 PRINT "FOR A MINIMUM R OF 5K C MUST BE ";A;" MICRO FARADS"
90 PRINT "FOR A MAXIMUM R OF ";B;"K C MUST BE ";C;"MICRO FARADS"
100 PRINT
110 PRINT "TYPE A VALUE OF R IN K OHMS TO OBTAIN THE VALUE OF C"
120 INPUT D
130 IF D<5 THEN PRINT "****WARNING VALUE OF R IS TOO SMALL****"
140 IF D>B THEN PRINT "****WARNING VALUE OF R IS TOO BIG****"
150 PRINT "VALUE OF C IN MICRO FARADS IS ";T/(1.1*D)
160 GOTO 120
```

Lines 50, 60 and 70 are straightforward arithmetic expressions, giving C for the smallest value of R (5K), the maximum value of R and its corresponding C, respectively. In lines 130 and 140 we check that the resistance value is in the recommended range. If it isn't a warning is printed.

If we wanted to print a range of values of C, for values of R between S and F, then the following changes to the program should be made. The best way to understand this program is to print out J, J*I and S + J*I each time through the loop.

```
110 PRINT "STARTING VALUE OF R (IN K OHMS)";
120 INPUT S
130 PRINT "FINISHING VALUE OF R (IN K OHMS)";
140 INPUT F
150 PRINT "HOW MANY VALUES ";
160 INPUT N
170 I=(F-S)/(N-1)
180 FOR J=0 TO N-1
190 PRINT "FOR R= ";S+J*I;"K OHMS C IS ";T/(1.1*(S+J*I));"MICRO FARADS"
200 NEXT J
```

Self-test questions

1 Write a program to print a conversion table from inches to millimetres covering the range 1 inch to 5 inches in one-tenth of an inch steps. Use the fact that 1 inch is approximately 25 mm.

2 Change the conversion program in question 1 so that the user can specify the range and the step size.

3 The actual resistance of a resistor changes with temperature according to the formula

$$R = R0(1 + aT)$$

where R0 is its resistance at 0 degrees C, a is the temperature coefficient and T is the working temperature. Write a program that will print the actual resistance over a range of temperatures. Allow the user to specify the resistance at zero and the temperature coefficient.

4 One of the untidy features of the 555 timer program is the way it always prints the results in kilohms when sometimes megohms would be more suitable (e.g. 5200K is better as 5.2M). Can you find a way of always printing the resistance in the correct units using only the BASIC statements covered so far?

5 Can you find another way of writing the FOR loop in the modification to the 555 timer program, using STEP?

4

Arrays and strings

Things really only start to get interesting in BASIC when you've mastered the use of arrays and strings. If you read an advanced book on programming you learn that a program is a little more than our earlier definition of 'a collection of BASIC statements' but is made up of two parts. One part roughly corresponds to our definition and is simply a collection of 'instructions'. The other part is usually referred to as the 'data structures'. Data structures are the things that our program does things to. In this sense we can liken the data structures to 'objects' or 'nouns' and the BASIC instructions to 'actions' or 'verbs'.

So far the only data structures we have are simple variables and thus the only programs that we have been able to write have also been pretty simple. The array is the first new data structure that we shall examine — for two reasons. First it's fairly easy to understand and second it's implemented in much the same way in all versions of BASIC. After looking at arrays we will then deal with the only other sort of data structure that BASIC has — the string.

Arrays

Roughly speaking an array is a collection of simple variables. If we have an array of, say, 10 variables then it's natural to talk about the first variable, the second and so on. It is this natural order of variables which is the essence of an array. To make life easy, in BASIC all the variables in an array have the same name. This is often referred to as the name of the array. So, for example, we might call our collection of 10 variables the array Z. The first variable we would call Z(1), the second Z(2) and so on to Z(10). Notice that we are talking about 10 distinct variables, Z(1) to Z(10). If by accident you refer to Z(11) or

Z(−1) most BASICs will tell you that you're being silly by giving you an error message. An array variable can be used anywhere that an ordinary variable can. The power of the array comes from the fact that the number that we use to specify which array variable we mean can be any arithmetic expression. For example (Z(I), Z(A+3), Z(2*A+1) are all valid.

The DIM statement

We still need to know how to tell the computer (so that it can reserve sufficient memory space) that an array is of a particular size. This problem is solved by the DIM statement. (DIM stands for DIMension.) The format of a simple DIM statement is

```
DIM variable name(number of array variables)
```

For example, our 10-variable array Z could have been declared in line 10 by

```
10 DIM Z(10)
```

Most versions of BASIC allow you to specify more than one array per DIM statement, e.g. DIM Z(10),A(25), but we will stick to one per statement for simplicity.

Other things to remember about DIM statements are that you must declare an array *before* you use it, and you cannot declare the same array more than once. Some BASICs relax these conditions (for example, ZX81 BASIC will allow you to redefine an array as long as you can accept the loss of the data in the old array) but it's safer to stick to them. One thing that some BASICs will let you do that's too good to miss is to allow the size of an array to be specified by an arithmetic expression. For example, DIM Z(N) or DIM Z(N+36) are both acceptable in enough BASICs to make it almost standard.

In general programming, arrays come in handy for two uses, the more common of which is storing a collection of numbers for later use. A lesser-known use is as a 'look-up' or 'code' table. We will give an example of storing and sorting data first.

Bubble sort

Suppose we have a set of readings and we want them sorted into descending order. We might use the program at the top of the next page.

This program is interesting for a number of reasons. The first part of the program from lines 10 to 70 is concerned with getting the readings into the array. Notice that lines 10 and 20 supply the *number* of readings, which is used to set up an array just big enough to hold all the

```
10 PRINT "HOW MANY READINGS";
20 INPUT N
30 DIM D(N)
40 FOR I=1 TO N
50 PRINT "READING NUMBER ";I;" = ";
60 INPUT D(I)
70 NEXT I
100 F=0
110 FOR I=1 TO N-1
120 IF D(I) > D(I+1) THEN GOTO 170
130 T=D(I)
140 D(I) = D(I+1)
150 D(I+1) = T
160 F=1
170 NEXT I
180 IF F=1 THEN GOTO 100
190 PRINT "READINGS SORTED"
```

readings in line 30. The FOR loop, lines 40, 50, 60 and 70, INPUTs the readings into each variable of the array in turn. Notice the use of I in the PRINT statement to say which reading is required and its use in line 60 in the INPUT statement.

The second half of the program, lines 100 to 190, sorts the readings using a method known as a bubble sort. A bubble sort works by examining each pair of variables in turn. If the first variable is bigger than the second nothing happens. However, if the first variable is smaller than the second they are in the wrong order (remember that we are aiming to end up with the largest readings coming first in the array). The answer is simple — just swap their positions. If you keep on doing this, starting at the top of the list and swapping the contents of variables that are the 'wrong' way round, you will eventually end up with the readings in the array in the correct order. You know that you have finished when you have passed right down the array, examining each pair, and have made no swaps because they are all in order.

Lines 110 to 170 scan through the array repeatedly. At line 120 we compare the two array variables D(I) and D(I+1). Notice how the ability to use arithmetic expressions to refer to array variables lets us call one array variable D(I) and the next D(I+1). If the order of the two readings is correct then no swap occurs and we GOTO 170, the end of the FOR loop. If a swap is needed, lines 130, 140 and 150 look after this. First D(I) is saved in a temporary variable T. Then D(I) is set equal to D(I+1) and D(I+1) is set equal to T. If we had not used the temporary variable to save D(I) we would have lost its reading when it was set equal to D(I+1). So swapping isn't so easy!

Another interesting use of a variable is F. To know when we've finished sorting we need to detect that no swaps have occurred during the FOR loop. If we set F equal to zero before we enter the FOR loop and set it to one if a swap occurs (in line 160), we can use it to test in line 180 for 'end of sort'. If F = 1 then a swap has occurred and we need to scan the array again, so we GOTO 100; if F = 0 then no swaps have occurred and the array is in order. When a variable is used in this way to indicate that something has or hasn't happened it is usually called a 'flag'.

The program isn't really finished. I've left the printing of the sorted array as an exercise. One last word on bubble sorting is that it is *not* the best way to sort things. You can find much quicker methods in standard books on programming but none that is quite as easy to understand.

Look-up tables

We are all familiar with look-up tables. If you're not so good at arithmetic then a written set of multiplication tables is a help. This is one of the simplest forms of the look-up table. To illustrate the point let's write a program for a look-up table for multiplication by two!

```
10 DIM M(12)
20 FOR I = 1 TO 12
30 M(I) = 2*I
40 NEXT I
50 PRINT "2 TIMES ";
60 INPUT J
70 PRINT " = ";M(J)
80 GOTO 50
```

Although this program isn't very useful it does show what you can do with a look-up table. Lines 10 to 40 load the table with the correct values. Lines 50 to 80 carry out the look-up. If you find this program difficult to understand just remember that the answer for 2*J is stored in M(J).

There are two reasons why we might use a look-up table rather than calculate the answer. First, a look-up table is often quicker than calculating the result each time. For example, instead of the two-times table we might look up sine or cosine. Second, there might not be a formula for the results stored in the table. For example, there is no formula for the number of dots and dashes in morse code symbols for each of the digits one to ten, and so a look-up table comes into its own for this application. And it is often the case that data on electronic components is given in the form of a graph; the only way that this can be used is by storing the values in a look-up table.

Most beginners have trouble with look-up tables when the 'input' to the table is not a simple whole number. For example, in the case of a look-up table for sines the angle would vary from 0 to 90 degrees but the look-up would have to be done in terms of 1 to N. The solution to this sort of problem is not too difficult. We simply 'chop up' the range of the input into N pieces. In the sine case, we might choose N = 9 and look up M(1) if the angle was between 0 and 10, M(2) if it was between 10 and 20, and so on, until we reach M(9) with the angle between 80 and 90. The changing of the angle into whole numbers 1 to 9 can be done either by a series of IF statements or by a single calculation:

```
J = INT(X/10) + 1
```

where X is the angle and INT is a new BASIC function to change a real number into a whole number (an INTeger) by chopping off the fractional part (e.g. INT(3.445) = 3). Notice that you have to be careful when using this method because if X = 89 then J = 9, but if X = 90 then J = 10, which is larger than the size of the array. In other words, be careful about end points!

Many dimensions
Most of the look-up tables that we use are a little more complicated than the ones we've been considering so far. For example, if you want to know the road distance between two towns, the table would have two inputs, the name of the first town and the name of the second town. If we look at the table it is obvious that it spreads out in two directions, one corresponding to the starting town and one to the finishing town. The table is said to be 'two-dimensional'.

We can extend this idea to BASIC arrays. A two-dimensional BASIC array is a collection of variables all with the same name and ordered by two numbers. Examples of two-dimensional array elements are D(3,3) and M(1,2). A two-dimensional array is defined using the DIM statement

```
DIM array name(maximum array size 1,maximum array size 2)
```

For example, the 10 by 10 variable array, B, would be declared by

```
DIM B(10,10)
```

If the array D is declared to be M by N, i.e. DIM D(M,N), the total number of variables that we have created is M*N which can be rather more than you'd expect. The innocent little array B of size 10 by 10 uses 100 variables. Starting with M(1,1) and ending with M(10,10) you can list 100 different variables — try it. The point is that a two-dimensional array can very quickly use up all your storage space.

Some computers (the ZX81 for instance) will allow you to define arrays of three, four, five or more dimensions. These are not so often used as one- and two-dimensional arrays and eat up storage even faster.

A resistor selector

As a practical example the following program will select a suitable resistor from the standard range of preferred values. When you work out the value of a resistor that you need in a particular circuit you usually end up with an ugly-looking number like 45943 ohms. The trouble is that you can't buy resistors of 45943 ohms. The resistor manufacturers only produce certain values — the so-called E12 and E24 ranges. The E12 range of values is such that for any resistance value you want there is an E12 preferred value that is within plus or minus 10 per cent of it. Similarly, the E24 are within 5 per cent of any value. Clever, isn't it?

The program given below asks for a calculated value and then checks the E12 table to find the closest preferred value. It prints the value along with the percentage error that you are making by using it. (Notice that this percentage error must be less than 10 per cent by definition of the E12 series.)

```
10 PRINT "RESISTOR SELECTOR"
20 DIM R(13)
30 R(1)=1
40 R(2)=1.2
50 R(3)=1.5
60 R(4)=1.8
70 R(5)=2.2
80 R(6)=2.7
90 R(7)=3.3
100 R(8)=3.9
110 R(9)=4.7
120 R(10)=5.6
130 R(11)=6.8
140 R(12)=8.2
144 R(13)=10
145 S=1
150 PRINT "CALCULATED RESISTOR VALUE IN OHMS = ";
160 INPUT V
170 IF V<10 THEN GOTO 210
180 V=V/10
190 S=S*10
200 GOTO 170
210 J=1
220 M=R(1)-V
230 M=ABS(M)
240 FOR I=2 TO 13
```

```
250 D=R(I)-V
260 D=ABS(D)
270 IF D<M THEN J=I
280 IF D<M THEN M=D
290 NEXT I
300 PRINT "NEAREST PREFERRED VALUE = ";R(J)*S
310 PRINT "PERCENTAGE ERROR = ";(V-R(J))/V*100;" % "
320 GOTO 145
```

If you have followed this book so far, you'll find this program quite straightforward so I'll leave it to you to improve it. Two possible improvements are to make it work for the E24 series as well, and if there are two best values to print out both. The only interesting points are the scaling of V to a number less than 10 in lines 170 to 200, and the way S is used to remember how many tens we've divided by. Lines 210 to 290 find the smallest difference between V and the table of E12 values. In lines 230 and 260 we use another new BASIC function ABS to find the absolute difference. The function ABS simply throws away the sign and makes all numbers positive. If you study this program carefully the rest should be obvious. See if you can answer the following questions:

(a) Why do we set J = 1 in line 210?
(b) What does line 270 do?
(c) Would the program still work if we swapped lines 270 and 280?

Strings

When we come to look at strings we have something of a problem. It's not that strings are any more difficult to understand than arrays, it's just that their implementation differs from BASIC to BASIC. The main trouble is that three very popular BASICs do things with strings in *very* different ways. Microsoft BASIC (on micros like Apple, Pet, etc. — for more detail see Chapter 1), ZX Spectrum BASIC and ZX81 BASIC all go their own way with strings. Of the three, Microsoft BASIC is the most widely accepted and is considered to be the standard BASIC. Because of this we will deal with strings in the way Microsoft does, but for the benefit of the ZX81 and ZX Spectrum community an explanation of how to convert from Microsoft will be given.

What is a string?

We have already come across strings in PRINT statements. For instance:

```
PRINT "This is a literal string"
```

As explained in Chapter 2, the characters between the quotes form a literal string. A string is nothing more than a collection of characters — a 'string' of characters. Most BASICs demand that strings are enclosed by quotes and it is important to realise that the quotes are not included in the string.

Obviously, what we are lacking at the moment is any way of saving strings for later use. We need variables that can hold strings — a string variable. Most BASICs use a $ sign to indicate that a variable is capable of holding a string. For example, A$ is a string variable, just as A is a numeric variable. You can use both A and A$ in the same program because they mean completely different things.

String expressions

As with numerical variables, we can assign a value to a string variable. The only difference is that the value is a string.

For example:

```
10 A$ = "MY NAME"
```

assigns the string "MY NAME" to the variable A$. But if you try

```
A = "MY NAME"
```

you'll get an error message because A without the dollar sign is, by definition, a numeric variable. Similarly

```
A$ = 234
```

will give you an error because the string is not enclosed within quotes. You can only assign a string to a string variable.

```
A$ = "234"
```

is acceptable though, because "234" *is* a string.

If we add the line

```
20 PRINT A$
```

to our example and run the resulting program we see

```
MY NAME
```

printed on the screen. Notice that the quotes have vanished. Remember, the quotes are *not* part of the string. They are used to show where it starts and ends.

A string variable can be used in INPUT and PRINT statements in the same way as a numeric variable. For example:

```
10 INPUT A$
20 PRINT "STRING =", A$
```

is perfectly correct.

We can now save strings for later use but at the moment we have no way to alter strings. The simplest method of altering strings goes under the very extravagant name of 'concatenation', which simply means joining two strings together. The standard symbol for concatenation is + but this should not be taken to mean that it has anything to do with addition. Concatenation is easier to understand via an example:

```
10 PRINT "WHAT IS YOUR SURNAME";
20 INPUT S$
30 PRINT "WHAT IS YOUR FIRST NAME";
40 INPUT F$
50 N$ = F$ + S$
60 PRINT "YOUR NAME IS ";N$
```

If you run this program you will be asked for your surname and first name and the program will print out your name the right way round. The only trouble is that it won't leave a space between the two. Most BASICs will allow you to get over this problem by

```
50 N$ = F$ + " " + S$
```

But be careful because some BASICs will only allow one + per statement. In such versions you would have to use

```
50 N$ = F$ + " "
51 N$ = N$ + S$
```

String IFs

So far strings haven't been of much practical use to us — all we can do is input them, print them and add them together. Most of the time strings are used to save the answer to a question and then to decide what the answer means. For example, consider the following:

```
10 PRINT "DO YOU WANT TO RUN THIS PROGRAM (YES = 1,NO
= 0)";
20 INPUT A
30 IF A = 0 THEN STOP
```

The computer asks its question in English but cannot handle anything but a number as an answer. It would be very much better if the answer could be YES or NO. This is possible with the use of strings in IF statements. For example:

```
10 PRINT "DO YOU WANT TO RUN THIS PROGRAM (ANSWER YES
OR NO)";
20 INPUT A$
30 IF A$ = "NO" THEN STOP
```

The only two relations that can be tested are equality, e.g. A$="NO", and inequality, e.g. A$<>"YES". The reason for this is that the 'greater than' and 'less than' relationships have no standard meaning when applied to strings — some BASICs will let you use them but what they do varies. (Later we will see that > and < can be used on strings of only one letter and that we can compare the lengths of strings.) In the example above the answer NO will stop the program and the answer YES will continue it, but so will the answer Y, STOP, OK or any other string you might type in. The trouble is that we test for the answer NO but not for YES. The line

```
40 IF A$<>"YES" THEN GOTO 10
```

needs to be added in order to do this.

String functions

As we can add strings together using the plus sign it might seem that all we have to do is use a symbol such as the minus sign to mean 'take one string from another', but strings are not like numbers. The sort of thing that we might want to do is to take the first few characters from one string and place them in another. This is difficult to write in the same way as arithmetic so we use string functions rather than symbols. String functions are similar to mathematical functions such as SIN(X). SIN(X) takes the number X and works out the corresponding sine, which is also a number. A string function takes a string and works out an answer which is also a string. As we said, the trouble with string functions is that they are not standard so we will deal with the Microsoft versions first.

```
LEFT$(A$,n)
```

returns the n leftmost characters of the string A$. For example,

```
PRINT LEFT$("ABCDE",3)
```

prints ABC.

```
RIGHT$(A$,n)
```

returns the n rightmost characters of the string A$. For example,

```
PRINT RIGHT$("ABCDE",2)
```

prints DE.

```
MID$(A$,n,m)
```

returns n characters, starting at the mth. For example,

```
PRINT MID$("ABCDE",2,2)
```

prints BC.

There are other string functions that can be found in BASIC but these are the most common and the most useful.

There is one other function concerning strings that we need to know about. It differs from the three above in that it does *not* return a string as its answer but a number.

```
LEN(A$)
```

returns the number of characters in the string A$. For example,

```
PRINT LEN("ABCDE")
```

prints 5.

Using string functions

The best way to make sense of the string functions is to use them. The most common difficulty that new programmers have with string functions is using them to carry out simple operations that need the use of more than one of them. For example, how would you print out a string letter by letter? Try:

```
10 INPUT S$
20 FOR I = 1 TO LEN(S$)
30 PRINT MID$(S$,I,1)
40 NEXT I
```

This method can be used to scan through a string looking for a particular character.

Another problem is how to print out a string minus the first n letters. Try:

```
10 INPUT S$
20 PRINT RIGHT$(S$,LEN(S$) – N)
```

To understand how this program works notice that if you don't want to print the first N characters then just print the last LEN(S$) – N characters!

A more difficult problem is to insert one string into another. Suppose we want to insert the string A$ into the string S$ after the Nth character

```
10 INPUT A$,S$,N
20 A$ = LEFT$(A$,N) + S$ + RIGHT$(A$,LEN(A$) – N)
30 PRINT A$
```

The expression in line 20 looks frightening! If you look at it carefully the first part extracts the first N characters and the last part extracts the remainder of the string. After breaking the string A$ down into these two parts the concatenation operator is used to put the parts back together again with S$ in the middle.

ZX81 and ZX Spectrum BASIC

As mentioned earlier the ZX81 and ZX Spectrum handle strings differently from Microsoft BASIC. In some ways Sinclair's methods are simpler and more powerful than RIGHT$, LEFT$ and MID$.

Rather than using string functions the ZX81 uses a notation called 'slicing' to describe the part of the string that we are interested in. The slicing notation is

```
string (N TO M)
```

and means 'the part of the string from character N to character M'. For example, "ABCDEF"(3 TO 5) is CDE. The only thing that makes the slicing notation a little difficult is the number of things you can leave out. If you leave out N then 1 is assumed, if you leave out M then the length of the string is assumed. You can also leave out the TO and the result is the same as if you'd written (N TO N). A few examples might help:

```
"ABCDEF" (  TO 3) = "ABCDEF" (1 TO 3) = "ABC"
"ABCDEF" (3 TO  ) = "ABCDEF" (3 TO 6) = "CDEF"
"ABCDEF" (  TO  ) = "ABCDEF" (1 TO 6) = "ABCDEF"
"ABCDEF" (3)      = "ABCDEF" (3 TO 3) = "C"
```

The only other things to remember are that N and M must refer to a part of the string that exists unless N is larger than M. For example "ABC" (2 TO 4) gives an error but "ABC" (4 TO 2) returns the null string (explained later).

The slicing equivalents of the Microsoft string functions should now be easy for the reader to construct. For LEFT$(A$,N) use A$ (1 TO N) or A$ (TO N), for RIGHT$(A$,N) use A$ (LEN(A$)−N TO) and for MID$(A$,N,M) use A$ (N TO N+M) — it's really quite easy.

Using these methods it is easy to convert any program using the Microsoft functions into slicing notation. However, the ZX81 has more to offer in that not only can the part of a string specified by a slicer be extracted and used, it can also be assigned to. For example

```
10 A$ = "ABCDEF"
20 A$(2 TO 4) = "******"
30 PRINT A$
```

prints A***EF. Notice that the part of the string specified by the slicer (2 TO 4) has been replaced by the string of asterisks and that only the first three asterisks have been used.

Atom BASIC

In general, the Atom does things so differently that the best advice to Atom users is to read their manuals carefully. The only danger is that programming in Atom BASIC will result in isolation from the rest of the BASIC community. This can be avoided by reading the rest of this book and becoming familiar with the Microsoft approach!

The Atom's method of handling strings is a little difficult to explain in the light of what we have learned so far. Essentially the difference is in the way the name of a string is used. In Atom BASIC the name of a string is used as a 'pointer' to the beginning of the string. The Atom also demands that you tell it the maximum size each string will have before you use it. In short, the differences are:

(a) Strings are referred to by names like $A rather than A$.
(b) To define the maximum length of each string you have to dimension an array of the same name.
(c) The + operation doesn't work; instead of A$ = B$ + C$ use $A = $B $A + LEN(A) = $C.
(d) For A$ = RIGHT$(B$,N) use $A = $B + LEN(B) − N.
(e) For A$ = LEFT$(B$,N) use $A = $B $A + N = "".
(f) For A$ = MID$(B$,N,M) use $A = $B + N $A + M = "".
(g) For LEN(A$) use LEN(A).

The null string

This is an appropriate time to introduce a very special string — the null string. The null string plays the role of zero for string functions and compares, and is a string containing no characters. For example A$ = "" is the null string and should not be confused with B$ = " "

which is a string containing a single blank. The difference is also clear in that LEN(A$) is zero and LEN(B$) is one.

Numbers to strings and strings to numbers
Earlier it was pointed out that A$ = 234 is not allowed because 234 is not a string. What if we want to use a number as part of a string? The solution is to use the function STR$(X) which returns the string that would be produced if X were printed. For example

```
10 X = 345.56
20 A$ = STR$(X)
30 PRINT A$,X
```

will print 345.56 twice. This may not seem very useful at the moment but for advanced programming it can be invaluable. For example, if you want to print a number only if it will fit into a specific amount of space, you could use

```
10 IF LEN(STR$(X)) < 4 THEN PRINT X
```

Going the other way, from a string of numbers to a value, is just as easy. The function VAL(A$) returns the numeric value of the string A$. For example, X = VAL("234") puts two hundred and thirty-four into the variable X.

Single characters and their order
If you were asked for the first letter of the alphabet you would have no trouble in giving the answer A. In the same way that you order the alphabet, your computer orders all of the symbols that it can print. Unfortunately this order is not the same for all computers so we have to be a little careful. A very useful function is CHR$(N) which returns the character that comes Nth in your computer's order of things. If you want to know what order your machine uses, try this program:

```
10 FOR I = 0 TO 255
20 PRINT "CHARACTER ";I,"IS",CHR$(I)
30 FOR J = 1 TO 300
40 NEXT J
50 NEXT I
```

which will print out each character in turn. The FOR J loop in lines 30 to 50 simply wastes time so that you can see each character in turn. If you see some unexpected spaces going up the screen, remember that some of the characters will be unprintable — things that you cannot see, such as blank and backspace.

Now that we know the order that the machine uses we can return to an earlier topic. If we want to sort some names into order we have to be able to compare where letters come in that order. This can be done by using the 'greater than' and 'less than' relations in an IF statement. For example

```
10 INPUT A$
20 INPUT B$
30 IF A$<B$ THEN PRINT A$;" COMES BEFORE ";B$
40 IF B$<A$ THEN PRINT B$;" COMES BEFORE ";A$
```

If you type Z in answer to the first INPUT and then X in answer to the second INPUT then the program (should) print out X COMES BEFORE Z. The 'less than' relationship for single-character strings is better read as 'comes before' and the 'greater than' is better read as 'comes after'. The extension to 'less than/greater than or equal to' should be obvious. What happens if you try < or > with strings of more than one character depends on the version of BASIC you have. It is best avoided but if you want to know what happens you can use the program above to find out how your BASIC handles this problem.

String arrays

As we discovered earlier, arrays are very useful and it is nice to know that we can have arrays of strings as well as numbers. The idea of a string array is very similar to that of an ordinary array. We need to tell the computer how many elements the string array has and this is achieved via the DIM statement

```
DIM string variable(number of elements)
```

For example, an array of 10 strings could be declared using

```
10 DIM A$(10)
```

A simple example will help explain string arrays. Suppose that we want to save a list of favourite colours.

```
10 DIM C$(5)
20 FOR I = 1 TO 5
30 PRINT "NAME A FAVOURITE COLOUR OF YOURS";
40 INPUT C$(I)
50 NEXT I
60 PRINT "SO YOUR FAVOURITE COLOURS ARE -"
70 FOR I = 1 TO 5
80 PRINT I;"...";C$(I)
90 NEXT I
```

Notice that an important feature of a string array is that there is an implied order of the elements. That is, if we want to know the first colour then we look at C$(1).

Some machines will even allow you to have multi-dimensional string arrays and there is not much new to say about these except that you don't often need them, and it's a good job because they eat up memory at an alarming rate!

Morse code trainer

In these days of advanced electronics it is surprising to find that the morse code is still in use. It has to be admitted that under difficult conditions a simple morse signal will get through where any sort of speech signal will at best be uncertain. Morse or CW transmitters are also very simple and take very little bandwidth. After all these advantages it should come as no surprise to discover that one of the conditions of being granted an amateur radio licence (Class A UK) is a reasonable proficiency at morse.

What follows, the largest program example in this book so far, is a morse code tutor. It makes extensive use of strings but should be easy to convert to almost any machine.

```
10 DIM C$(26)
20 C$(1)=". -"
30 C$(2)="- . . ."
40 C$(3)="- . - ."
50 C$(4)="- . ."
60 C$(5)="."
70 C$(6)=". . - ."
80 C$(7)="- - ."
90 C$(8)=". . . ."
100 C$(9)=". ."
110 C$(10)=". - - -"
120 C$(11)="- . -"
130 C$(12)=". - . ."
140 C$(13)="- -"
150 C$(14)="- ."
160 C$(15)="- - -"
170 C$(16)=". - - ."
180 C$(17)="- - . -"
190 C$(18)=". - ."
200 C$(19)=". . ."
210 C$(20)="-"
220 C$(21)=". . -"
230 C$(22)=". . . -"
240 C$(23)=". - -"
250 C$(24)="- . . -"
260 C$(25)="- . - -"
270 C$(26)="- - . ."
300 PRINT "A MORSE SYMBOL WILL APPEAR - PRESS THE CORRESPONDING KEY"
310 PRINT "YOU WILL BE GIVEN THREE GOES TO GET IT RIGHT BEFORE"
320 PRINT "THE COMPUTER WILL GIVE YOU THE CORRECT ANSWER"
330 PRINT
340 PRINT "ARE YOU READY (YES/NO)"
350 INPUT A$
360 IF A$="NO" THEN STOP
370 IF A$<>"YES" THEN GOTO 300
380 FOR I=1 TO 10
390 C=INT(RND(0)*25+1)
400 J=0
```

```
410 J=J+1
420 PRINT "WHAT IS    ";C$(C)
430 INPUT A$
440 IF A$=CHR$(64+C) THEN GOTO 500
450 PRINT "****WRONG***   TRY AGAIN"
460 IF J<3 THEN GOTO 410
470 PRINT "THE SYMBOL STANDS FOR ";CHR$(64+C)
480 NEXT I
490 GOTO 530
500 PRINT "CORRECT !"
510 S=S+1
520 NEXT I
530 PRINT
540 PRINT "YOUR SCORE IS ";S;"/10"
550 IF S<5 THEN PRINT "KEEP TRYING"
560 IF S>4 THEN PRINT "VERY GOOD"
570 IF S>7 THEN PRINT "YOU CAN'T DO MUCH BETTER!"
580 IF S=10 THEN PRINT "YOU HAVE FULL MARKS!"
```

Lines 10 through to 270 set up an array C$ containing the morse symbols for A to Z. The symbol for A is stored in C$(1), that for B is stored in C$(2), and so on, down to Z in C$(26). Lines 300 to 330 print some instructions (feel free to improve on these) and lines 340 to 370 ask if the user is ready to start. Lines 380 to 530 is the part of the program that does most of the work. Ten morse symbols are selected at random. The FOR loop starting at 380 is responsible for repeating everything ten times and the statement at line 390 selects one symbol at random.

Line 390 needs some explanation. RND(0) is a function which generates random numbers in the range 0 to 1. You'll have to check that RND(0) works on your machine because it's not always standard. Once you have obtained a number between 0 and 1, multiply it by 25 to get a number between 0 and 25. If you then add 1 you obtain a number between 1 and 26. The function INT is used to convert this number into an integer, i.e. it chops off any fractional part. For example, INT(3.14159) is simply 3. It should now be clear what line 390 is doing; it returns a random whole number in the range 1 to 26. This number is stored in C and is used to select which morse symbol will be displayed in line 420.

J is zeroed in line 400 and is used to count how many guesses the user has had. The user's answer is accepted in line 430 and is compared with the correct answer in line 440. The only difficult thing about line 440 is the use of CHR$(64+C). On most machines the alphabet starts after 64 other characters, i.e. A is the 65th character. In our array of morse symbols, C$, the symbol for A is stored in C$(1) so to generate A using CHR$(I) I must be 65, i.e. 64 + 1, and in general to produce the character corresponding to the code stored in C$(I) we use CHR$(64+I). This obviously will not work on all computers because it depends on the order in which characters are produced by CHR$(I). For example, CHR$(38) produces A on the ZX81 so we would have to use CHR$(37+I) in our morse program.

If the answer given was wrong we end up at lines 450 and 460, which print out a message and check to see if the user has had more than three guesses. If there have been fewer than three guesses then the question is repeated by going to line 410. If there have been more than three guesses then the answer is printed and another symbol is selected at random by the action of the NEXT I instruction. If the FOR loop is completed, i.e. ten symbols have been presented, then line 490 is carried out taking us to line 530.

If the answer supplied was correct line 500 prints a message and line 510 adds one to the score stored in S. Again a new symbol is selected by the action of the NEXT I and if the FOR loop is complete then control passes to 530. The lines 530 through to 580 simply print out how well the user has done and are easy enough to understand. However, notice that more than one of the congratulatory messages can be produced for any one score.

I can think of many improvements to this program but I leave them all to you! One thing that would make the program more realistic, however, would be to use whatever sounds your machine can produce actually to signal the morse code rather than just display it.

Self-test questions

1 Write a program that will read in 10 numbers and then print them out in reverse order.

2 Change the program in question 1 so that it will reverse any number of numbers.

3 Write a program that will input a string and then print it out in reverse order.

4 Write a program that will accept the colour coding on a resistor and print out its value. The colour code is:

Colour	*Number*
black	0
brown	1
red	2
orange	3
yellow	4
green	5
blue	6
violet	7
grey	8
white	9

The first two colours on the resistor give the first and second most significant figures and the third gives the number of zeros following the second most significant figure.

5

Functions, subroutines and expressions

Functions and subroutines are very important in BASIC because, properly used, they can make programs shorter and easier to debug. In some ways a subroutine is an extended form of a function so it makes sense to look at functions first.

Functions

In BASIC there are two distinct forms of function — user-supplied and built-in functions. The built-in functions are the better known of the two types and indeed we have been making use of them without making too much fuss in earlier chapters. A function is, roughly speaking, a single command that causes some calculation to be carried out and returns or supplies only *one* answer. For example, SQR(4) causes the computer to find the square root of 4. The single answer is of course, in this case, 2. As only a single answer is produced by a function, we can use the equals sign to say where this answer should be stored for later use. For example:

```
10 A = SQR(4)
20 PRINT A
```

would print 2. Notice that if a function returned more than one value as an answer, it would be impossible to use the equals sign without inventing a way of selecting which of the possible answers was to be stored in the variable. For this reason the function SQR only supplies the positive square root of the number and totally ignores the negative one. So while we know that there are two square roots of 4 (+2 and −2) this fact is for humans only!

Another result of the one-answer property of functions is that they can be used in arithmetic expressions without much restriction. For example:

```
10 A = (SQR(4)*3) + 2
20 PRINT A
```

would print 8.

The rule is that BASIC will evaluate a function (work out the answer) and then continue to evaluate the rest of the expression as if the function were replaced by its answer. So (SQR(4)*3) becomes (2*3) and then 6. In general, functions have a higher priority than any other arithmetic operations and are worked out first.

The value or values that a function is given to work on are often referred to as the argument or the parameters of the function. For example, in SQR(X) X is the argument. It is not necessary for a function to have only one argument or parameter although most do. One other luxury that most BASICs allow is the use of expressions inside a function. That is, parameters may be arithmetic expressions. For example:

```
10 A = SQR(2*2)
20 PRINT A
```

will print 2. If you're feeling daring, try:

```
10 PRINT SQR(SQR(SQR(2)))
```

The type of answer a function can give is not limited to numbers; string functions, for instance, return strings. Similarly the parameters of a function can be of different types.

Most BASICs are supplied with a bewildering range of functions. It's true to say that the better the BASIC the more the functions. Some functions that are worth knowing about include:

ABS(X) returns the absolute value of X, i.e. it chops off any negative sign. ABS(2) is 2; ABS(−2) is also 2.

COS(X) returns the cosine of X. The only thing to be careful of is the units that X is measured in. All the BASICs that I know of require X to be in radians. To convert degrees to radians multiply by 180 and divide by 3.14159.

EXP(X) returns e raised to the power of X. EXP becomes very large very quickly and the largest value of X that can be used is often not much more than 80.

INT(X) returns the largest integer smaller than or equal to X. INT(1.5) is 1 but INT(−1.5) is −2 because −2 is *smaller* than −1.

LOG(X) returns the log (base e) of X.

RND(X) returns a random number between 0 and 1. The effect of the value of X varies quite a lot from BASIC to BASIC. RND is discussed more fully in the next chapter.

SIN(X) returns the sine of X. (See the notes on COS.)

SQR(X) returns the positive square root of X. The value of X must be positive.

TAB(X) is a very special function because instead of returning a value it causes the next character to be printed in column X unless the current print line has already gone beyond that column. Most BASICs only allow TAB to be used in a PRINT statement.

TAN(X) returns the tangent of X. If you haven't got TAN then don't forget that tan(X) = sin(X)/cos(X).

There are of course many functions that are not listed here (all of the string functions for example) but you should be able to look up the more exotic ones in your machine's BASIC manual.

As an example of using supplied functions the following program plots sine curves (a subject that is covered more fully in the next chapter).

```
10 X = 0
20 N = 20
30 PRINT TAB(INT((SIN(X) + 1)*N));"*"
40 X = X + 0.5
50 GOTO 30
```

Although this program is very short it's well worth studying. Line 30 is the most complicated involving three functions. SIN(X) returns a value between plus and minus one. Adding one and multiplying by N changes this into a number between zero and 2*N. The function INT then changes this into a whole number. TAB causes the * to be printed in the appropriate column. This method can be used for plotting any graph or function with values that lie between, or can be converted to lie between, 0 and 2*N.

User-defined functions

Let's suppose that we want to find the roots of a quadratic equation. For example, consider:

$ax^2 + bx + c = 0$

If you remember the formula that gives you the solution it's easy to solve. The first root is given in BASIC by

```
10 R1 = (-B + SQR(B*B - 4*A*C))/(2*A)
```

and the second root is given by

```
20 R2 = (-B - SQR(B*B - 4*A*C))/(2*A)
```

Life would be a lot easier if there was a BASIC function called ROOT that supplied the first root of a quadratic equation. For instance,

```
10 R1 = ROOT(A,B,C)
```

Unfortunately, BASIC isn't quite that clever but it will allow you to define a function that will do the same job. (I only know one version of BASIC that will not let you define a function and that is ZX81 BASIC. However, some versions have restrictions on what sort of function you can define; Dragon or Tandy Color BASIC, for example, will only let you use one parameter.) To define a function we use the DEF FN instruction:

```
DEF FNx(parameter list) = function definition
```

The x may be replaced by a valid variable name, in our case any single letter; the parameter list is simply a list of variables separated by commas and the function definition is any arithmetic expression that will fit on one line. For example, if we declare the function FNR as the first root of a quadratic as

```
10 DEF FNR(A,B,C) = (-B + SQR(B*B - 4*A*C))/(2*A)
```

then the first root of any quadratic can be found by adding the line

```
20 PRINT FNR(A,B,C)
```

It is important to note that the letters or names used for the parameter list have nothing to do with any variables of the same name used elsewhere in the program. Once defined, a user-supplied function can be used exactly like a built-in function. In particular any variables can be used as parameters (not just those named in the parameter list) and the function can be used as many times as desired. For example, assuming the function definition FNR has already been defined earlier in the program, all of the following lines are correct:

```
10 A = FNR(A,B,C)
20 A = FNR(P,C,D)
30 PRINT FNR(X,Y,Z)*FNR(T,R,F) + 3
```

A more advanced point about the use of parameters in functions is that any variables used in the function definition that are not named in the parameter list *are* taken to be the same as variables used in the main program. For example:

```
10 DEF FNA(N) = N*V
20 V = 2
30 PRINT FNA(3)
```

will print 6. The parameter N is set equal to 3 when using the function in line 30, but the variable V is set equal to 2 by line 20 and is not changed by the use of the function.

As an indication of how functions can be used in a more inventive way, consider the problem of defining a function to find both roots of a quadratic. It may seem impossible, because a function can return only one answer at a time, but examine the following lines of program:

```
10 DEF FNR(A,B,C,N) = (−B + N*SQR(B*B − 4*A*C))/(2*A)
20 PRINT "TYPE THE THREE COEFFICIENTS";
30 INPUT A,B,C
40 PRINT "THE FIRST ROOT IS ";FNR(A,B,C,1)
50 PRINT "THE SECOND ROOT IS";FNR(A,B,C,−1)
```

The trick is to include an extra parameter to indicate which root has to be calculated, i.e. N. If N is 1 then the first root is calculated and if it is −1 the second root is produced.

This example also serves to show up the major limitation of BASIC functions. It is possible for there to be no real roots of a quadratic equation. This condition shows itself by the expression inside the SQR function evaluating to a negative number. As we all know the SQR function cannot work with negative numbers and this causes the program to print an error message and stop. The solution to this problem would be to include a test (i.e. an IF statement) that bypasses the SQR function in case of trouble. But of course this is impossible because function definitions can only include *one* line of BASIC.

Subroutines

In BASIC subroutines are not as well developed and useful as the facilities provided by other languages but they are still worth knowing.

about. A subroutine is a chunk of program that performs some task that is often used. For example, if during a program we need to zero an array a number of times, instead of using

```
10 FOR I = 1 TO N
20 A(I) = 0
30 NEXT I
```

each time, it would be nice if we could use a single line to somehow invoke the same little bit of program. This could be done by writing the little bit of program starting at a high line number, e.g. 1000, and then jumping to it using a GOTO every time it was required. The trouble with this method is that of getting back to the point in the program that follows the appropriate GOTO. If you don't believe me, try it! To make this easier BASIC has two extra statements:

GOSUB line number

and

RETURN

GOSUB works like a GOTO, in that control is passed to the line number indicated, but in addition the line number that follows the GOSUB is stored for future use. The RETURN statement finds the line number stored by the last GOSUB and transfers control to it. An example will make this clear:

```
10 DIM A(20)
20 GOSUB 1000
30 PRINT "ARRAY ZEROED"
40 A(3) = 3
50 GOSUB 1000
60 PRINT "ARRAY ZEROED AGAIN"
70 STOP
1000 FOR I = 1 TO 20
1010 A(I) = 0
1020 NEXT I
1030 RETURN
```

At line 20 control is passed to line 1000 and the array is zeroed. The RETURN statement in line 1030 of the subroutine causes control to be passed back to line 30, the line following the most recent GOSUB. At line 50 the subroutine is called again and RETURNs to line 60.

Other points to notice about subroutines are:

(a) A subroutine may call another subroutine and the RETURN statement will take care of where to transfer control at the end of each.

(b) A RETURN without a GOSUB having been encountered will cause an error.
(c) All of the variables within a subroutine are the *same* variables used in the rest of the program (including other subroutines).

The use of subroutines to replace frequently used chunks of program is not the only way in which subroutines can be employed. In fact the use of subroutines has been developed into a philosophy, or even a science, of programming method known as 'Top Down Structured Programming', one of the main ideas of which is the almost obsessive use of subroutines at every opportunity. The result of this is that a program consists mainly of a long list of subroutine calls making it easier to alter and understand at a later date. If you want to write good programs then the rule is to use a subroutine unless you can think of a good reason not to! You will find examples of this style of programming in the larger programs given later in this book.

Expressions

One of the first subjects that we looked at was the way that arithmetic was performed in BASIC. This was then extended to include arithmetic expressions that involved variables, and now functions. The truth of the matter is that when any information is changed in BASIC the agent of change is an expression. For example, in the previous chapter, which discussed the use of strings, string concatenation and string functions were used to form string expressions. Later we will discover that there is one other form of expression in BASIC — the logical expression. In short, an expression is a sort of recipe for combining and altering data of any sort. If the data happens to be numbers we have an arithmetic expression, if the data are characters we have a string expression and, finally, if the data are logical values (true or false) we have a logical expression. Notice that an expression contains the idea of doing something. This is the main difference between an arithmetic expression and an algebraic formula. The formula may represent some truth, that one quantity always equals another or the way that quantities are related, but a BASIC arithmetic expression is something that the computer is obliged to work out.

This idea of working out an expression, or evaluating it, is worth looking at a little closer. If the computer evaluates an arithmetic expression, say, then the result is a number; and why should this number be treated any differently from any other number that you have used in a BASIC program? Why, for example, shouldn't you be able to write things like

```
PRINT 2*SQR(3) + 5
IF A + 34>E/2 THEN . . .
FOR I = 55 + J TO N – 1 STEP SIN(.5)
```

The answer is that in most versions of BASIC you can. The general rule in BASIC is that anywhere that you can use a number you can use an arithmetic expression and vice versa. Similarly, anywhere that you can use a string you can use a string expression and finally, as we shall see later, anywhere that you can use a logical value you can use a logical expression. So far these freedoms have been introduced as special cases. When the FOR loop was introduced it was explicitly stated that you could use arithmetic expressions for the starting value, final value and step size. However, this is just an example of the general principle outlined above. Some versions of BASIC take this principle to its logical conclusion and allow you to write commands such as GOTO X*3 + 34, but this is not to be encouraged as it can make programs very difficult to debug.

Self-test questions

1 Write a program that will print a table of positive square roots between any given values and with a given step size.

2 Change the program in question 1 to print a table of sine and cosine values.

3 If your version of BASIC has user-defined functions write such a function to return the log to the base 10 of a number using

$$\log_{10} x = \log_e x \log_{10} e$$

given that $\log_{10} e = 0.434295$

4 Write a program that will evaluate

S = x – x ↑ 3/6 + x ↑ 5/120 – x ↑ 7/5040

for a given value of x. Use either a subroutine or a user-defined function actually to work out the result. You might like to compare the results that you get with SIN(x) where x is measured in radians.

6

Using BASIC — 1

At this point we have enough BASIC to tackle some quite major projects. However, electronics is a subject that involves the use of some techniques, such as complex numbers, that BASIC was never designed to cope with. In this chapter we will look at some of the ideas that characterise the use of BASIC in electronics via some simple examples.

Plotting waves

In the last chapter a program was given and discussed that plotted a continuous sine wave using the TAB function. Although many small computers have very good graphics facilities this sort of crude plot is often to be preferred. There are many reasons for this. The complete lack of standardisation of graphics commands makes writing programs that make any use of graphics very machine-specific. Also, a graphics printer is still the exception rather than the rule and if you cannot print out a graph its use is very restricted. Finally, the sort of graphs that are of use in electronics do not always need to be drawn with the very high accuracy of a modern graphics system. Very often it is only the overall shape and the values of a few critical points that interest us.

As you already know, one of the central ideas of electronics is that any waveform can be produced by adding together nothing but sine (and cosine) waves. Although this is a well-known theory it is always much better to see theory in action. The following short program will plot the sum of a series of sine waves.

This program has been written using subroutines to carry out every major task. It is worth noticing how this use of subroutines makes the program much easier to understand. The main program, lines 10 to

```
10 K=0
20 L=40
30 DIM Y(50)
40 INPUT A
50 INPUT B
60 GOSUB 1000
70 GOSUB 2000
80 GOSUB 3000
90 K=K+1
100 GOTO 40

1000 X=1
1010 FOR T=0 TO 2*3.14159 STEP 2*3.14159/50
1020 Y(X)=Y(X)+A*SIN(K*T)+B*COS(K*T)
1030 X=X+1
1040 NEXT T
1050 RETURN

2000 M=Y(1)
2005 S=Y(1)
2010 FOR I=1 TO 50
2020 IF Y(I)>M THEN M=Y(I)
2030 IF Y(I)<S THEN S=Y(I)
2040 NEXT I
2050 IF M=S THEN S=M/2+1
2060 RETURN

3000 FOR I=1 TO 50
3010 PRINT TAB((Y(I)-S)*L/(M-S)+.5);"*"
3020 NEXT I
3030 RETURN
```

100, consists of almost nothing other than a list of subroutine calls. Subroutine 1000 works out the effect of adding the next sine and cosine term in the series. The variable K is used to keep a count of the number of terms, and A and B are the amplitudes specified by the user for sine and cosine respectively. Subroutine 2000 finds the maximum and minimum values stored in the array Y. Finally, subroutine 3000 plots the graph using the TAB function as described in the previous chapter. The only difference is that this plot routine will work for any list of values stored in Y as it automatically scales the plot to fit the screen. This is done by using the largest and smallest values stored in M and S, and the number of characters to a screen line stored in L.

The main program is in the form of a loop and each time through the loop the user is asked for values of A and B and then a term of the form

A*SIN(K*T) + B*COS(K*T)

is added to the current values in Y. In this way the shape of the series can be viewed as it builds up to more and more terms of higher and higher frequency (notice that K increases by one each time through the loop).

As a demonstration of the program in action try entering the following values for A and B:

K	A	B
0	0	0
1	1	0
2	0	0
3	0.333	0
4	0	0
5	0.2	0
6	0	0
7	0.143	0

You should be able to see an increasingly good approximation to a square wave appear on the screen. If you want to carry the series on to more terms the rule is that even terms have zero values for A and B and odd terms have A equal to 1/K and B equal to zero.

Using nothing but very simple BASIC methods we have produced a program that will show the Fourier series at work. It's not just in theory that adding sine waves together is a way to produce any waveform!

Histograms

Another useful form of graphical display is the histogram. As long as you are prepared to accept a sideways-on display, a histogram is even easier to produce than a sine plot. Suppose we want to produce a histogram representing the data 6, 10, 3. All we have to do is print a line of six asterisks, then a line of ten and then three. That is:

```
I******
I**********
I***
```

You can probably guess that a program to do this will involve two FOR loops, one to print the required number of lines and one to print the required number of asterisks in each line. If the data of the previous example is stored in the array H then the following program will produce the histogram:

```
1000 FOR I = 1 TO 3
1010 PRINT "I";
1020 FOR J = 1 TO H(I)
1030 PRINT "*";
1040 NEXT J
1050 PRINT
1060 NEXT I
```

The outer FOR loop starting at line 1000 will print 3 lines. The inner FOR loop starting at line 1020 will print H(I) asterisks. The only other point worth mentioning is the way that the semicolon is used in lines 1010 and 1030. If you end a PRINT statement with a semicolon then any subsequent PRINT statements carry on printing where the first one left off. In other words the semicolon suppresses the normal starting of a new line that follows a PRINT statement.

Although the program contains the basic method of plotting histograms, it is hardly useful. In particular, it cannot cope with a wide range of data without plotting off the screen. The answer to this problem is to search through the array H to find the largest value and use this to scale the histogram to fit on the screen. This scaling is simple enough: if the largest value is M (for max) and the number of asterisks in the largest line is L then the number of asterisks that should be printed for H(I) is

```
INT(H(I)*L/M)
```

where INT takes the integer part of the result to avoid trying to print a fractional number of asterisks. The final program is as follows.

```
10 DIM H(50)
20 L=40
30 PRINT "HOW MANY VALUES ";
40 INPUT N
50 FOR I=1 TO N
60 PRINT "VALUE ";I;" = ";
70 INPUT H(I)
80 NEXT I
90 GOSUB 1000
100 GOTO 30

1000 GOSUB 2000
1010 FOR I=1 TO N
1020 PRINT "I";
1030 A=INT(H(I)*L/M)
```

```
1040 IF A=0 THEN GOTO 1080
1050 FOR J=1 TO A
1060 PRINT "*";
1070 NEXT J
1080 PRINT
1090 NEXT I
1100 RETURN

2000 M=H(1)
2010 FOR I=1 TO N
2020 IF M<H(I) THEN M=H(I)
2030 NEXT I
2040 RETURN
```

The main program is concerned with reading the data into H and then calling subroutine 1000 to plot a histogram. Subroutine 1000 is essentially what we had before except that it now calls subroutine 2000 to find the maximum value in H and then uses the result to scale the histogram as described earlier. The only other point of interest is the use of the IF statement in line 1040 to skip the inner FOR loop when there are no asterisks to be printed.

Random numbers

The BASIC function RND was introduced in the previous chapter but without very much explanation of how to use it. The main problem in using the RND function is exactly how to write it. Some versions of BASIC use RND, others use RND(0) and yet others RND(−1), all to mean the same thing. The RND function will produce random numbers between 0 and up to but not including 1, and it's up to you to find out the exact form of the function that your BASIC uses. Once you have found out try

```
10 PRINT RND
20 GOTO 10
```

to check that you are getting random numbers between 0 and 1.

It is not often that random numbers between 0 and 1 are required so it is worth knowing how to change their range into something more generally useful. If you need random integers between A and B, then use:

```
INT(RND*(B − A + 1)) + A
```

If you need fractional numbers between A and B then use:

```
RND*(B – A) + A
```

The main use of random numbers in electronics is in the production of a representative range of test data. For example, instead of typing in values to fill the array H to test the plotting program given above, we could generate random numbers. That is, change lines 60 and 70 to

```
60 H(I) = INT(RND*100)
70 PRINT "VALUE ";I;" = ";H(I)
```

which will fill the array with numbers from 0 to 99.

The main feature of the numbers generated by RND is that (at least in theory) they are all equally likely to turn up. Thus, in the last example any of the numbers in the range 0 to 99 had the same probability of being produced. It is very often the case that we want to examine the way slight variations away from an ideal component value will affect the performance of a circuit and then what we really want to generate are numbers of the form 'ideal value + error', where 'error' is the random factor altering the ideal value. It is nearly always the case that the error term has a normal distribution rather than the even distribution of numbers produced by RND. You can produce numbers with a normal distribution of standard deviation S (mean zero) by using

```
SQR(2)*(RND+RND+RND+RND+RND+RND–3)*S
```

This is only an approximation but it is good enough for most purposes.

Complex arithmetic

Very few versions of BASIC provide any facilities for handling complex numbers. However, if you deal with the real and imaginary parts separately it is very easy to produce a few subroutines, or user-defined functions, for complex addition, subtraction, multiplication and division.

If your version of BASIC allows two-letter names for variables then it is a good idea to create two variables for each complex variable. The first would have a name ending in R for the real part and the second would have a name ending in I for the imaginary part. For example, FR and FI could be used to hold the real and imaginary parts of a complex frequency.

Complex addition and subtraction are easy to deal with. If w and v are complex numbers then

re(w+v) = re(w) + re(v)
im(w+v) = im(w) + im(v)

and

re(w−v) = re(w) − re(v)
im(w−v) = im(w) − im(v)

where re stands for the real part and im for the imaginary part. In other words, to add two complex numbers all you have to do is add the real and imaginary parts separately; and similarly for subtraction the real and imaginary parts can be dealt with separately. This is so simple that there is really no need to write special functions or subroutines for addition or subtraction. However, when it comes to multiplication and division things are a little different. If v and w are complex numbers then

re(v*w) = re(v)*re(w) − im(v)*im(w)
im(v*w) = re(v)*im(w) + im(v)*re(w)

and

$$re(v/w) = \frac{re(v)*re(w) + im(v)*im(w)}{re(w)*re(w) + im(w)*im(w)}$$

$$im(v/w) = \frac{im(v)*re(w) - re(v)*im(w)}{re(w)*re(w) - im(w)*im(w)}$$

If your version for BASIC supports user-defined functions with four parameters then you can use

```
DEF FNM(VR,VI,WR,WI) = VR*WR − VI*WI
```

and

```
DEF FNN(VR,VI,WR,WI) = VR*WI + VI*WR
```

for the real and imaginary parts of multiplying VR,VI and WR,WI respectively and

```
DEF FND(VR,VI,WR,WI) = (VR*RW + VI*WI)/(WR*WR + WI*WI)
```

and

```
DEF FNE(VR,VI,WR,WI) = (VI*WR − VR*WI)/(WR*WR + WI*WI)
```

for the real and imaginary parts of dividing VR,VI by WR,WI. If you can't use four-parameter user-defined functions then you will either have to write out the expressions as and when they are required or use subroutines instead.

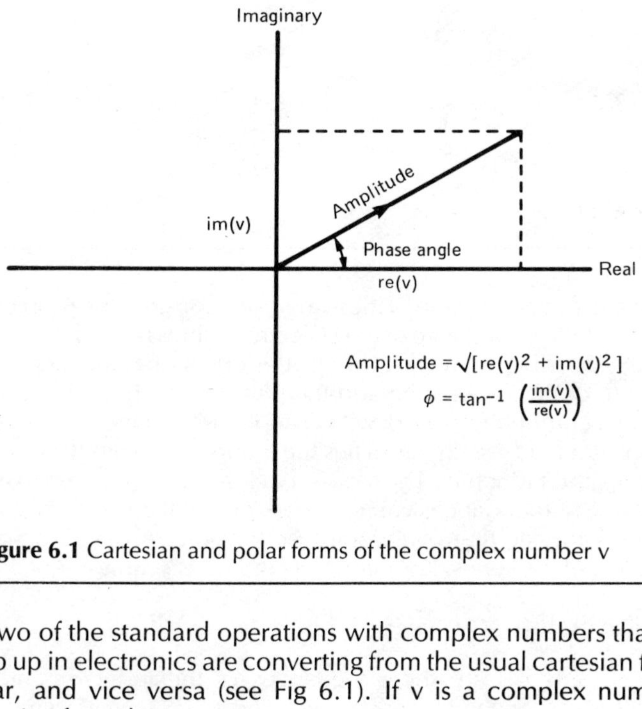

Figure 6.1 Cartesian and polar forms of the complex number v

Two of the standard operations with complex numbers that often crop up in electronics are converting from the usual cartesian form to polar, and vice versa (see Fig 6.1). If v is a complex number in cartesian form then

r = SQR(re(v) ↑ 2 + im(v) ↑ 2)
t = ARCTAN(im(v)/re(v))

where r is the modulus and t is the argument (or phase angle). Most versions of BASIC include the ARCTAN function but it is sometimes called ATAN or even ATN so check your BASIC manual. Converting from polar to cartesian is even easier:

re(v) = r*COS(t)

and

im(v) = r*SIN(t)

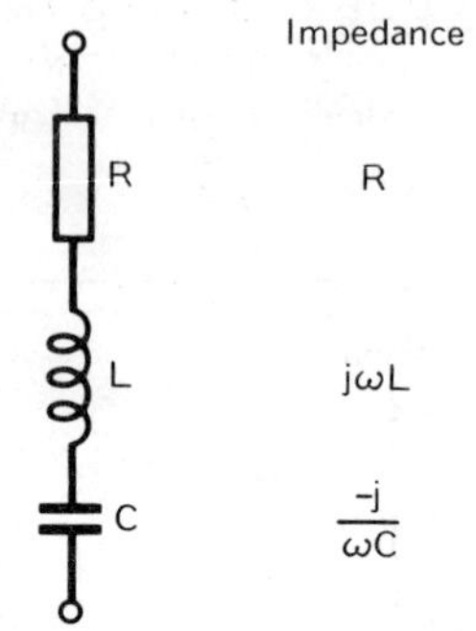

Figure 6.2 LCR series circuit

As an illustration of some of these methods consider the problem of working out the impedance of the LCR circuit shown in Fig. 6.2. Now it is quite possible to work through the equations and derive the magnitude of the impedance and the phase angle but it is easier, if you have a computer, to work with complex impedances. The use of complex numbers greatly simplifies the theory of circuits that involve capacitors and inductors. The idea is that if you are prepared to work with complex impedances you can work out total impedances using the same formulae that you would use for pure resistances. So the impedance of the series LCR circuit of Fig. 6.2 is simply

$$Z = R + ZL + ZC$$

where ZL and ZC are the impedances of the inductor and the capacitor respectively. Using $ZL = j\omega L$ and $ZC = -j/(\omega C)$ gives

$$Z = R + j(\omega L - 1/(\omega C))$$

where ω is the angular frequency. Now, to investigate the response of the LCR circuit, all we have to do is to work out this equation for various values of ω. To make any sense of the results, however, it is necessary to convert the complex impedances to polar form and then plot two graphs, one of impedance and one of phase angle. These processes are carried out by the following program.

Subroutine 1000 works out the complex impedance for a wide range of frequencies and stores the real part in ZR and the imaginary part in ZI. If your version of BASIC doesn't support two-letter names then all you have to do is rename the arrays. Subroutines 2000 and 3000 plot the magnitude and the phase angle of the impedance using

```
10 DIM ZR(50),ZI(50)
20 S=40
30 L=.001
40 C=.000001
50 M=100
60 INPUT R
70 GOSUB 1000
80 GOSUB 2000
90 GOSUB 3000
100 GOTO 20

1000 K=1
1010 FOR F=1000 TO 20000 STEP 500
1020 ZR(K)=R
1030 ZI(K)=2*3.14159*F*L-1/(2*3.14159*F*C)
1040 K=K+1
1050 NEXT F
1060 RETURN

2000 FOR I=1 TO 25
2010 A=SQR(ZR(I)*ZR(I)+ZI(I)*ZI(I))
2020 PRINT TAB(A*S/M);"*"
2030 NEXT I
2040 RETURN

3000 FOR I=1 TO 25
3010 A=ATN(ZI(I)/ZR(I))+2*3.14159
3020 PRINT TAB(A*S/4/3.14159);"X"
3030 NEXT I
3040 RETURN
```

techniques that have been well described in earlier sections. The program as it stands only allows the user to alter the value of R so as to examine its damping effect on the resonance. The resonance itself can be clearly seen as a dip in the magnitude plot and in the phase change in the phase plot.

You can use this program to experiment with the effect of changing the values of L and C, but notice that the scaling for the plots is set in line 50 and this will almost certainly have to be adjusted to keep the plot on the screen. Even if you are very familiar with the mathematics of the LCR series circuit there is a lot to be learned by 'playing' with this program. For example, the resonant frequency is set by the product of L and C so what is the effect on the response of keeping the product fixed but changing the proportions of L and C?

You can investigate the response of similar combinations of L, C and R using this program. All you have to remember is to work out the total impedance just as if all the components were pure resistors and then substitute $j\omega L$ and $-j/\omega C$ for inductive impedances and capacitive impedances respectively. With a computer, passive AC circuits are easy to understand.

Programming

This chapter has presented a number of simple programs both to illustrate programming and to suggest the sort of attitude that you should have toward the computer. None of the programs could be considered complete in any sense and you should take up the challenge of changing them into finished, and therefore useful, products. You will learn a lot about BASIC by reading and improving on other people's programs and it is important to get into this habit right from the start. If you don't then you will often have to rewrite programs that already exist just because they don't quite do what you want them to.

The computer is a powerful tool to enable you to explore the calculations that lie behind electronics. You must never be afraid to experiment and by building up a collection of subroutines and techniques of the sort that have been presented in this chapter you can even reduce the amount of work involved in writing programs.

7

Logic

Many people believe that computers and computer programmers spend all of their time dealing with logic. If you've read the earlier chapters you'll know that this is far from the truth. To write programs in BASIC, or any other computer language, does not require any understanding of traditional logic. For the subject of digital electronics, however, things are very different. The design of any digital equipment requires at least an intuitive understanding of logic. BASIC can help with digital electronics because it can evaluate logical expressions in the same way that it can evaluate arithmetic expressions. After learning about logic in BASIC it is easy to find many other uses for it apart from helping with digital design — so while programming can help with electronics the practice in logic helps the programming!

Boolean logic

There are many different types of logic but the one that concerns us here is one of the simplest — Boolean logic. Boolean logic is based on the simple fact that a statement is either true or false. For Boolean logic to work all that matters is that it must be one or the other, i.e. true or false are the only possibilities.

Consider the BASIC statement A>0. Does this say 'the variable A *is* greater than zero'? Suppose A is to set equal to one earlier in the program then A>0 is true, but if we set A equal to minus one then A>0 is false. It should be clear now that A>0 doesn't mean that we are saying A *is* greater than zero but that it is a statement which may be true or false. In this sense A>0 is not a statement of fact but an expression to be evaluated (see Chapter 5). A statement which may be true or false is known as a Boolean expression. You could use an expression like A>0 in a line like this, for example:

```
10 IF A>0 THEN PRINT "A IS BIGGER THAN ZERO"
```

We have been using Boolean expressions within IF statements throughout this book. The general form of the IF statement can now be expressed as

IF Boolean expression is true THEN statement

The Boolean expressions that we already know are all based on the relational operators introduced when we dealt with the IF statement. These are:

=	equal to	A=B
<	less than	A<B
>	greater than	A>B
<>	not equal to	A<>B
<=	less than or equal to	A<=B
>=	greater than or equal to	A>=B

Our new understanding of these means that each is a statement which may either be true or false.

Notice that the equals sign has two meanings in BASIC. It can be used to mean assignment, i.e. LET A = 0, or it can be used as a relational operator, i.e. IF A = 0 THEN etc. In the first case A *is* set equal to 0; in the second A is unchanged and instead a true or false result is returned. Also remember that the relational operators can be used to compare strings as well as numbers — see Chapter 4.

Boolean operators

So far Boolean expressions have given us nothing but a re-interpretation of old facts. If this was all there was to logic it would hardly be worth the bother! But consider the following short program:

```
10 PRINT "INPUT TWO RESISTANCE VALUES";
20 INPUT R
30 INPUT S
40 IF R<0 THEN PRINT "RESISTANCE MUST BE BIGGER THAN
   ZERO!"
50 IF S<0 THEN PRINT "RESISTANCE MUST BE BIGGER THAN
   ZERO!"
     (rest of program)
```

At lines 40 and 50 we test to see if the two resistance values are smaller than zero. If you examine these two lines carefully you might think that there is some repetition because the same thing happens if either resistor is less than zero, and if both are less than zero the message is printed twice. As a simple exercise the reader might like to try to change the program to print the message just once no matter what.

From a commonsense point of view it would be nice if we could write in place of lines 40 and 50:

```
40 IF R<0 OR S<0 THEN PRINT "RESISTANCE MUST BE BIGGER
   THAN ZERO"
```

In fact the part of the statement between the IF and the THEN is still a Boolean expression; that is, it can either be true or false. To see that this is the case we can examine what makes the IF statement carry out the statement following the THEN.

R<0	S<0	THEN taken
False	False	No
False	True	Yes
True	False	Yes
True	True	Yes

By examining this table it should be clear that the course of action following THEN is taken if either of the two inequalities is true; and this is of course what we mean by OR. Going back to the definition of the IF statement the Boolean expression R<0 OR S<0 must be true if the statement following the THEN is carried out. We can now write the earlier table as:

R<0	S<0	R<0 OR S<0
False	False	False
False	True	True
True	False	True
True	True	True

This is called a 'truth table' and it is one of the basic ways of understanding logic expressions. Apart from understanding a logic table we have also discovered our first logical operator — OR.

It is worth pausing for a moment to think about the similarities between what we have just done and the more familiar evaluation of arithmetic expressions. In arithmetic we write A + B and everyone understands that for various values of A and B we could work out the result of A + B. We could write down an 'arithmetic table' listing all the values that A and B could have and the result A + B. This would take rather a long time however as the number of values that A and B can take is infinite! In the case of logical expressions we don't have the same trouble as each variable can only take on two values — true or false — and we can write the truth table for A OR B very easily.

Logical variables

Going back to our analogy with arithmetic, the two values, true and false, are logical constants in the same way that 0, 1, 2, 3, 4, etc. are arithmetic constants. It would be nice if BASIC provided us with logical variables that we could set equal to logical constants. For example:

```
10 A = TRUE
20 IF A THEN PRINT "A TRUE"
```

You may find this difficult to read and it may look a little strange, but you should be able to understand what it's getting at. We are using the variable A as a logical variable (i.e. it can only store the values true/false) and at line 10 it is set equal to the constant true. At line 20 the IF statement evaluates the Boolean expression A and finds that it is true so the action specified by THEN is taken.

The trouble is that most BASICs do not have logical variables so this explanation is not exactly correct. It is however correct in spirit, as we shall see. Although the two values that a Boolean expression can take on are usually called true and false, these names are entirely arbitrary. We could call them 'plink' and 'plonk' or anything else without altering Boolean logic at all. More to the point we could call them 0 and 1 and use ordinary arithmetic variables to hold logical values! This is the method used by most BASICs. The only extra complication is that not everyone can agree on 0 and 1 and some use 0 and −1, or −1 and +1, etc. The differences are irritating but not too important. A brief summary of how the different BASICs handle Boolean logic can be seen in the following table. (The 'Full evaluation range' column is explained on page 76.)

BASIC	*True*	*False*	*Full evaluation range*
Microsoft V5	0	−1	−32768 to 32767
ZX81/Spectrum	1	0	See below
Apple	1	0	Not applicable
BBC	−1	0	−65536 to 65535

For the remainder of this book we will assume that 1 is true and 0 is false. Any changes that are needed to make the programs work with another convention are fairly easy and provide good practice at working with logic. With this decided, our earlier program becomes

```
10 A = 1
20 IF A THEN PRINT "A TRUE"
```

It is now possible for us to use the computer to produce the truth table for OR:

```
10 FOR I = 0 TO 1
20 FOR J = 0 TO 1
30 PRINT I;"  ";J"  ";I OR J
40 NEXT J
50 NEXT I
```

Logical operators

We arrived at our first logical operator — OR — by appealing to common sense. It would be possible to continue to apply common sense and derive the other often used logical operators, but to save time and avoid tedium the following BASIC program will print truth tables for OR, AND and NOT:

```
10 PRINT "TRUTH TABLE FOR AND–OR–NOT"
20 FOR I = 0 TO 1
30 FOR J = 0 TO 1
40 PRINT I;"  ";J;"  ";I AND J;"  ";I OR J;"  ";NOT I
50 NEXT J
60 NEXT I
```

AND used between two logical variables will result in the expression being true only if both variables are true. NOT is a unary operator (see Chapter 2) and simply changes the truth value of a single variable into the other value; i.e. NOT changes true into false and false into true. We have already dealt with OR. These three basic logical operators are all we need to express any Boolean logic. At a more practical level AND, OR and NOT are all we need to express any set of conditions for an IF to carry out the statement following the THEN.

Logical operators can be combined into expressions in much the same way as arithmetic operators. (The only thing you have to be careful of is the priority of the operators. Most BASICs define the priorities so that, unless you use brackets, NOT is done first, then AND and finally OR.) For example NOT(I) AND I, NOT(A<0) AND (B<0), (A=0) OR ((A=1) AND (B=0)) are all valid logical expressions. The rules for manipulating such expressions are not difficult to learn and if you're interested there are many good books on simple logic that will help you become proficient in manipulating logic. For the purposes of programming, however, becoming an expert logician is more than we require — the computer can always work out any difficult logic for us. For example, if you need to know the truth table for NOT(I) AND I then change line 40 in the previous program to:

```
40 PRINT I;"  ";J;"  ";NOT(I) AND I
```

If you're lucky enough to own a ZX81 or ZX Spectrum or use a version of BASIC that has a full VAL or EVALuate instruction, try changing the previous program by:

```
10 PRINT A$
40 PRINT I;" ";J;" ";VAL(A$)
```

This will allow you to type in *any* logical expression of I and J and print out its truth table, because the function VAL works out the expression contained by the string A$. If you want to work out a truth table with three variables all you have to do is add a FOR loop with a variable K, another for four variables, yet another for five . . . until you run out of memory or patience.

Simplifying expressions

There is one thing that a computer cannot do very easily for you and that is to simplify Boolean expressions. Sometimes while working out what should go inside an IF statement you might come up with something like:

```
10 IF NOT(A<0) THEN . . .
```

You might have said to yourself, 'IF A isn't smaller than zero then . . .' and translated 'isn't smaller than zero' into BASIC as NOT(A<0). There is nothing wrong with this method of constructing IF statements — it's the way we all do it — but if you've been programming for a while you notice that NOT(A<0) is the same thing as A>=0. Put into words, 'A isn't smaller than zero' is the same as 'A is bigger than or equal to zero'. By making this simple change you can make the program slightly simpler and faster to run.

There are a whole host of such simplifications that you can make to a compex logical expression but it's up to you to decide if it's worth it. This is very similar to simplifying arithmetic expressions before using them in a program — its not wrong to use A*B + A*C rather than A*(B + C) but the latter is faster. We all learn how to simplify arithmetic expressions at school but logical expressions are something else! Fortunately we are saved by the fact that in practice the logical expressions that crop up are fairly simple and can be altered by using some common sense.

However, some rules for manipulating Boolean expressions might be useful:

Relations

1. NOT(A=B) = A<>B
2. NOT(A<B) = A>=B
3. NOT(A>B) = A=<B

Logic

1. NOT(NOT(A)) = A
2. A OR (B AND C) = (A OR B) AND (A OR C)
3. A AND (B OR C) = (A AND B) OR (A AND C)
4. NOT(A OR B) = NOT(A) AND NOT(B)

The rules for relations are easy enough to understand, but those for logic might be less obvious. The first logic rule says that to NOT something twice leaves it unchanged. The second and third rules are easier to understand if you compare them to arithmetic:

A*(B + C) = (A*B) + (A*C)

Thus both AND and OR can be factorised just like multiplication.

The final rule is the most difficult and is known as De Morgan's law. It is important because it provides the link between AND and OR, in that any expression involving OR can be changed into one involving AND and vice versa. This isn't a complete list of rules for manipulating logic but it will see you through most logical problems.

The other logical operators

AND, OR and NOT are all the logical operators that you *need* but sometimes it is useful to give names to often-used combinations of them. Moreover, not many BASICs offer more than AND, OR and NOT so it is important to know how to make up these more complicated operators.

NOT(A OR B) = A NOR B
NOT(A AND B) = A NAND B
(NOT(A) AND B) OR (A AND NOT(B)) = A EOR B

The first two examples are easy; the third, EOR, looks impossible! In fact EOR — Exclusive OR — is one of the most useful logical operators. To find out what it does, a truth table helps:

A	B	A EOR B
0	0	0
1	0	1
0	1	1
1	1	0

Looking at the last column you should be able to see a similarity with the truth table produced by OR. The difference is that EOR is true (i.e. 1) only when *one* of A or B is true, and OR is true if one *or both* are true.

Electronics and logic

At the start of this chapter it was explained that logic is used in the design (and debugging) of digital circuitry. This short section deals with the connection between programming and hardware logic.

Digital circuits are built up from a range of integrated circuits which provide elementary logic functions. For example, in the 74 range the 7400 contains four NAND gates. These provide the same operation as our NOT(A AND B) expression did in the last section, except that instead of 0 and 1 representing true and false two voltage levels (5 V and 0 V) are used and A and B are two inputs to the circuit. So if we have a network of simple logic integrated circuits we can write down a Boolean expression to represent them and use a program, based on the one given earlier, to work out a truth table and hence what outputs occur for any input. This is of course much easier than building the circuit.

This idea of translating a piece of simple digital electronics to a Boolean expression is the basis of a very powerful technique that can be used to try out any digital design. The reason why we have to extend the method is that most even slightly advanced digital equipment goes beyond 'static' logic by using oscillators and pulse generators to 'run' the logic circuits. So far we have not introduced a way of generating such pulses, or clocks as they are often called, but it's not too difficult and would make a very interesting project for the reader to tackle.

Evaluation range

This section is a little more advanced than the rest of the chapter so you may want to skip it until some later time. What happens if you try something like:

```
10 A = 6
20 B = 7
30 PRINT A AND B
```

So far we have only discussed logical operations involving values like 0 and 1. What happens if you try ANDing numbers like 6 and 7 depends on the version of BASIC that you are using. Some will treat the numbers as binary numbers (i.e. 6 = 110 and 7 = 111) and carry out the logical operation on each pair of bits in turn. Thus bit one of 6 is 0 and bit one of 7 is 1, so the result of ANDing them together is 0, and so on with the second and third pairs of bits, giving 110 as the answer. The result of this bitwise logical operation can of course be treated as another binary number and converted back to decimal — 110 is 6. Any BASIC that works this way will have a largest number that it can deal with; this was listed in the table given on page 72, in the column 'Full evaluation range'.

The ZX81 and ZX Spectrum are very special in their use of logical operations on extended ranges (i.e. other than 0 and 1). If you try:

```
10 PRINT A AND B
```

you will find that if B is 0 the result is zero, but if B is anything else the result is A. This remarkable feature also works for strings.

```
10 PRINT A$ AND B
```

will print nothing (i.e. the null string) if B is zero and A$ if B is any other value. This is particularly useful if you're trying to fit a program into a 1K ZX81 or a 16K ZX Spectrum because, instead of

```
10 IF B<>0 THEN PRINT A$
```

you can use

```
10 PRINT A$ AND B
```

The other logical operators on the ZX81 and ZX Spectrum are also special: A OR B is A if B is 0, and 1 if B is anything else; NOT A is 0 if A is not zero, and 1 if it is. Neither of these can be used with strings but they are very powerful statements and deserve further study. However, this comes under the heading of 'very clever programming' and is beyond the scope of this book!

Self-test questions

1 Work out the truth table for the following expressions

(a) A AND B AND C
(b) (A AND B) OR (A AND C)
(c) A AND (B OR C)

2 Write a single IF statement that will detect the fact that the value of a variable lies in the range L to H inclusive.

3 Write a program that calculates

Q = NOT(R AND NOT(Q AND S))

for all possible values of R, S and Q. Can you recognise the well-known digital circuit that this represents?

8

Using BASIC — 2

This further chapter is a collection of example programs written in BASIC. None of them is very long because that would defeat the object of presenting an example, but they are all finished in the sense that they do the job that was intended without any 'rough edges'. Unlike the programs in Chapter 6, therefore, they are examples of finished programs and they should be studied with this in mind.

Designing a Zener diode regulator

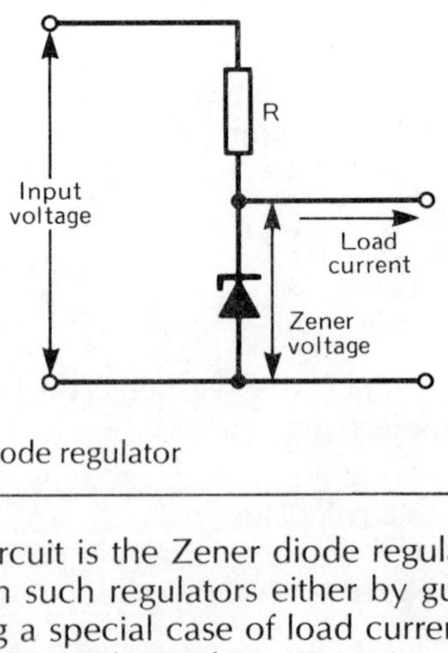

Figure 8.1 Zener diode regulator

A very common circuit is the Zener diode regulator (see Fig. 8.1). Most people design such regulators either by guessing the resistor value or calculating a special case of load current and hoping that everything will work at other values. To write a simple computer program to calculate a good value for R and tell you the power dissipation in both the resistor and the diode is very easy. If you know the minimum input voltage, the maximum load current and the Zener diode voltage, then a reasonable value for R is

$$R=\frac{\text{Max input V}-\text{Zener V}}{\text{Max load current}}$$

This equation is derived by considering the worst possible condition, i.e. when the load current is high and the supply voltage is low, and calculating the required resistance. (The voltage across the resistor is *always* the input voltage minus the Zener diode voltage.) There is one assumption in this calculation — that the Zener diode sinks no current! In practice the Zener diode does sink current and this should be added to the maximum load current. Because of this, the equation gives a value for R that is slightly high. Once we have the value for R the calculation of the worst case power dissipation is easy.

$$\text{Max resistor power}=\frac{(\text{Max input V}-\text{Zener V})^2}{R}$$

$$\text{Max Zener power}=$$

$$\text{Zener V}\times\left[\frac{(\text{Max input V}-\text{Zener V})}{R}-\text{Min load current}\right]$$

Using these simple design equations and what we have learned about BASIC it is not difficult to come up with the following Zener diode design program.

```
10 PRINT "MAXIMUM DC VOLTAGE";
20 INPUT V
30 PRINT "MINIMUM DC VOLTAGE";
40 INPUT W
50 PRINT "ZENER DIODE VOLTAGE";
60 INPUT Z
70 PRINT "MAXIMUM LOAD CURRENT";
80 INPUT J
90 PRINT "MINIMUM LOAD CURRENT";
100 INPUT I
110 LET R=(W-Z)/J
120 LET P=(V-Z)*(V-Z)/R
130 LET Q=((V-Z)/R-I)*Z
140 PRINT "APPROX RESISTOR VALUE","=";R
150 PRINT "MAX POWER IN RESISTOR","=";P
160 PRINT "MAX POWER IN ZENER","=";Q
```

Although this program is simple it is very useful and could be turned into a full Zener diode analysis program in no time at all — including features like allowing the user to try out a range of values for R and calculating the regulation factor. For example, the regulation factor (the ratio of the change in output volts for a change in input volts) is given by

1/(1 + R/RL + R/RZ)

where RL is the (fixed) load resistance and RZ is the Zener diode's slope resistance. Adding this on to the previous program gives the following.

```
170 PRINT "ZENER SLOPE RESISTANCE ";
180 INPUT S
190 A=Z/J
200 B=Z/I
210 PRINT "LOAD","REGULATION"
220 FOR K=B TO A STEP (A-B)/20
230 PRINT K,1/(1+R/K+R/S)
240 NEXT K
```

Notice the way that the range of load resistances is estimated using the maximum and minimum currents and the Zener diode voltage. Notice also the way that the table of stabilisation factors is given for 20 equally spaced intervals, no matter what the range of the load resistance.

The Wien bridge

For low frequency sine wave oscillators, the Wien bridge arrangement (see Fig. 8.2) is often used. The equation governing the impedance of the bridge itself is very easy to work out from Fig. 8.3 using complex impedances. The bridge itself is simply a potential divider with

$$Z1 = R1 - j/\omega C1 \qquad \text{(impedances in series)}$$

and

$$\frac{1}{Z2} = \frac{1}{R2} + j\omega C2 \qquad \text{(impedances in parallel)}$$

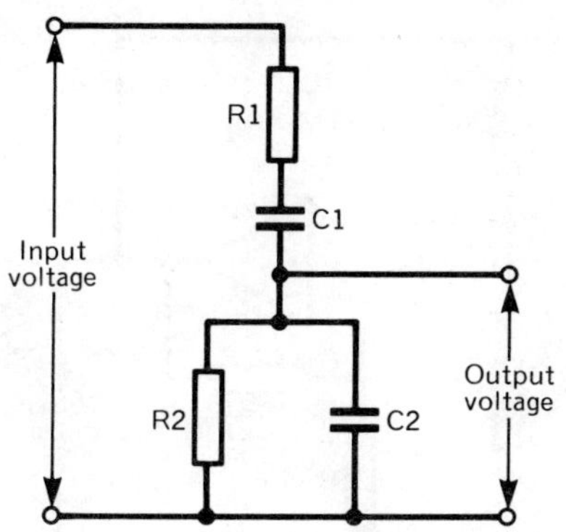

Figure 8.2 Wien bridge

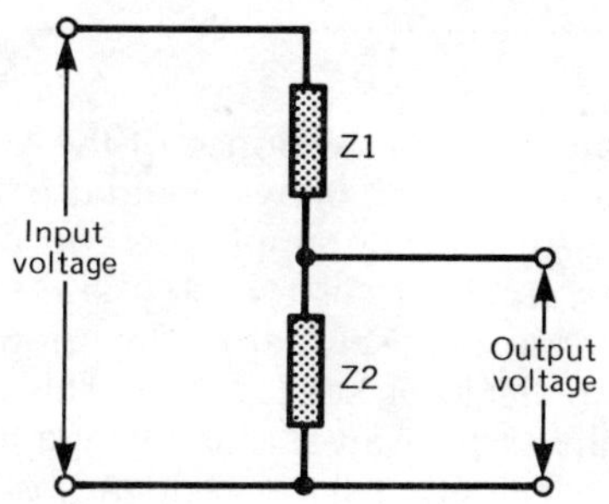

Figure 8.3 Equivalent potential divider

That is,

$$Z2 = \frac{R2}{1 + j\omega C2R2}$$

Using the usual formula for the voltage out of a potential divider we have

$$\frac{VI}{VO} = \frac{Z1 + Z2}{Z2}$$

$$= 1 + \frac{(R1+(1/j\omega C1))(1+j\omega C2R2)}{R2}$$

$$= 1 + \frac{R1}{R2} + \frac{C2}{C1} + j\left(\omega R1C2 - \frac{1}{\omega R2C1}\right)$$

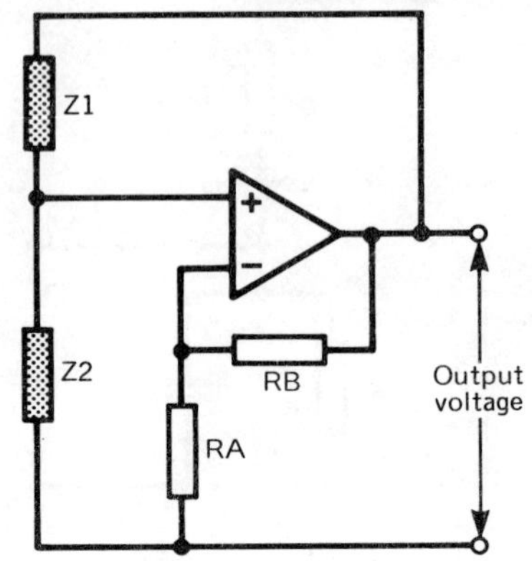

Figure 8.4 Wien bridge oscillator. RA and RB provide negative feedback to limit oscillation. RA is often a thermistor to improve stability

This last equation gives the response of the Wien bridge in the cartesian form. From this point on we could use the methods introduced in Chapter 6 to plot the amplitude and phase angle of the response. However, for an oscillator the phase shift introduced by the Wien bridge must be precisely zero. (The positive feedback effect comes from the amplifier itself, see Fig. 8.4.) This corresponds to the imaginary part of the response vanishing giving the required phase angle of zero. This implies that at the frequency of oscillation:

$$\omega R1C2 = \frac{1}{\omega R2C1}$$

In other words:

$$\omega = \frac{1}{\sqrt{(R1R2C1C2)}}$$

For computational simplicity it is generally assumed that R1 = R2 and C1 = C2, giving the frequency as

$$\omega = 1/RC$$

where ω is the angular frequency; $\omega = 2\pi f$.

Using this very simple equation we can work out the component values for any given frequency. It is usual to select a value of R and then work out the required value of C for the desired frequency. If this value of C is inconvenient or unobtainable then the value of R is altered and the calculation repeated. The following program makes this repetition easy.

```
10 PRINT "WIEN BRIDGE"
20 PRINT
30 PRINT "ALL RESISTORS IN K AND ALL CAPACITORS IN NF"
40 PRINT
50 PRINT
60 PRINT "RESONANT FREQUENCY IN HZ =";
70 INPUT F
80 PRINT "RESISTANCE VALUE =";
90 INPUT R
100 R=R*1000
110 C=1/(2*3.14159*R*F)
120 PRINT "CAPACITANCE REQUIRED = ";C*1E9;" NF"
130 GOTO 10
```

The only trouble with this design program is that it sticks to the traditional assumption that R1 = R2 and C1 = C2. In practice this is rarely the case because of component tolerances. A second problem is the difficulty of obtaining an exact frequency using preferred values for the components. Using a computer there seems to be no reason not to explore the possibility of using different values for each component. The following program will calculate the resonant frequency for a general Wien bridge.

```
10 PRINT "WIEN BRIDGE"
20 PRINT
30 PRINT "ALL RESISTORS IN K AND ALL CAPACITORS IN NF"
40 PRINT
50 PRINT
60 PRINT "R1= ";
70 INPUT R
80 R=R*1000
90 PRINT "R2= ";
100 INPUT S
110 S=S*1000
120 PRINT "C1= ";
130 INPUT C
140 C=C*1E-9
150 PRINT "C2= ";
160 INPUT D
170 D=D*1E-9
180 W=SQR(1/(R*S*C*D))
190 F=W/(2*3.14159)
200 PRINT "FREQUENCY = ";F;" HZ"
210 GOTO 10
```

Using this program in conjunction with the first it is possible to study the effect of slight changes in component values and the effect of selecting the next preferred value up and down. Thus the performance of the bridge can be optimised by trial and error.

This idea of using a computer as an aid to searching for a solution is fairly common in frequency design applications. It would be possible to write a much fuller program to find the optimum solution directly but, unless designing Wien bridges is a common activity, this would be more work than the application warranted.

Although the two programs given above serve well as aids to designing the Wien bridge you may be left with an uneasy feeling that there might be side-effects from not using a bridge in which R1 = R2 and C1 = C2. One way to examine any such side-effect is to examine the way that the frequency response of the bridge changes as component values are adjusted. We could use the sort of program given in Chapter 6 to plot the impedance but this is such a common requirement it is worth seeing if it is possible to do anything more general.

Plotting amplitude and phase

We already know one method of producing plots on the computer and it would seem that there is nothing more to add. However, in this next plot program the emphasis is on writing a general program that can be used to display any complex function in terms of amplitude and phase. Thus, any changes that have to be made to plot a different function should be kept isolated to individual subroutines.

In addition, to be of practical value the amplitude and phase should be displayed on the same plot and scales should be included. Believe it or not, it is this last requirement that is most difficult to fulfil. It takes a human to mark off any scale in a sensible fashion. Computers usually end up with fractional intervals and a silly number of divisions. For example, most people would divide a scale of 0 to 100 into 10 equal steps or, if this was too fine, five equal steps, but most automatic scaling routines are as likely to divide the range into six or four equal steps as anything else!

```
10 PRINT "FREQUENCY PLOTTER"
20 PRINT
30 GOSUB 1000
40 GOSUB 2000
50 GOSUB 3000
60 PRINT "NEW COMPONENT VALUES OR NEW SCALE FACTOR (C/S)";
70 INPUT A$
80 IF A$="C" THEN GOTO 30
90 IF A$="S" THEN GOTO 40
100 GOTO 60
```

```
1000 PRINT "R1= ";
1010 INPUT R
1020 R=R*1000
1030 PRINT "R2= ";
1040 INPUT S
1050 S=S*1000
1060 PRINT "C1= ";
1070 INPUT C
1080 C=C*1E-9
1090 PRINT "C2= ";
1100 INPUT D
1110 D=D*1E-9
1120 L=40
1130 RETURN

2000 PRINT
2010 PRINT "PLOT OVER FREQUENCY RANGE"
2020 PRINT "STARTING AT ";
2030 INPUT B
2040 PRINT "ENDING AT ";
2050 INPUT E
2060 IF B>E THEN GOTO 2000
2070 PRINT
2080 PRINT "SCALE FACTOR (HIGHEST IMPEDANCE)";
2090 INPUT M
2100 RETURN

3000 FOR I=0 TO M STEP M/10
3010 PRINT TAB(I*L/M+5);I;
3020 NEXT I
3030 PRINT
3040 FOR I=0 TO M STEP M/10
3050 PRINT TAB(I*L/M+6);"+";
3060 NEXT I
3070 PRINT
3080 FOR F=B TO E STEP (E-B)/20
3090 PRINT F;
3100 W=2*3.14159*F
3110 GOSUB 4000
3120 IF I>L THEN I=L
3130 IF J>L THEN J=L
3140 IF I<J THEN PRINT TAB(I);"X";TAB(J);"*"
3150 IF J<I THEN PRINT TAB(J);"*";TAB(I);"X"
3160 IF I=J THEN PRINT TAB(I);""
3170 NEXT F
3180 PRINT TAB(6);"+";
3190 PRINT TAB(L/2+6);"+";
3200 PRINT TAB(L+6);"+"
3210 PRINT TAB(5);"-PI";
3220 PRINT TAB(L/2+5);"0";
3230 PRINT TAB(L+5);"PI"
3240 RETURN

4000 GOSUB 9000
4010 A=SQR(ZR*ZR+ZI*ZI)
4020 P=ATN(ZI/ZR)
4030 I=INT(A*L/M+6.5)
4040 J=INT((3.14159+P)/2/3.14159*L+6.5)
4050 RETURN

9000 ZR=1+R/S+C/D
9010 ZI=W*R*D-1/(W*S*C)
9020 RETURN
```

A program to plot the frequency response is given above. The main program, lines 10 to 100, is in the form of loop which calls the appropriate subroutines as they are required. Subroutine 1000 is used to read in the component values. This would obviously have to be changed to plot a different equation. Subroutine 2000 reads in the range of frequencies over which the response is to be plotted and the scale factor, i.e. the highest value of impedance that will be plotted.

Subroutine 3000 is actually responsible for producing the plot. Lines 3000 to 3070 are concerned with producing the first horizontal scale corresponding to the amplitude. It is simply marked off in ten equal steps between 0 and M, the largest value to be plotted. Lines 3080 to 3170 produce the plot and call subroutine 4000 to return values of I and J which give the positions of the 'X' and the '*' respectively. The amplitude is plotted using the 'X' and the phase is plotted using the '*'. Notice that instead of subroutine 4000 returning actual values of amplitude and phase it scales them and returns the position on the line at which the symbols should appear. To plot both symbols on the same line we obviously have to print the one furthest to the left first as the TAB function cannot move the printing position backwards. This is achieved by lines 3140 and 3150. Line 3160 takes care of the case when the 'X' and the '*' fall in the same place, and lines 3120 and 3130 make sure that the line width is never exceeded. The final part of subroutine 3000 simply prints a second horizontal axis, this time marked up to show the phase.

Subroutine 4000 calculates the I and J from the real and imaginary parts of the value stored in ZR and ZI respectively. These values are calculated by subroutine 9000 from the formula for the frequency response. This and subroutine 1000 are the only two subroutines that have to be changed to plot a different frequency response.

Finally, the main program ends with a choice of either replotting the response with different component values or just with a changed scale factor. This allows the user to fix a scale factor by trial and error. Notice also that the user can obtain a neat set of axes by careful selection of the frequency range to be plotted.

The most important point about programming contained in this example is the way that subroutines make a program easier to write and understand.

Simultaneous equations and networks

Finding currents and voltages in networks that consist of resistors and batteries is simply a matter of applying Kirchhoff's law:

> The total current flowing out of a junction is equal to the total current flowing into the junction;

and the voltage law:

> The e.m.f. in a circuit equals the sum of the voltage drops around the circuit.

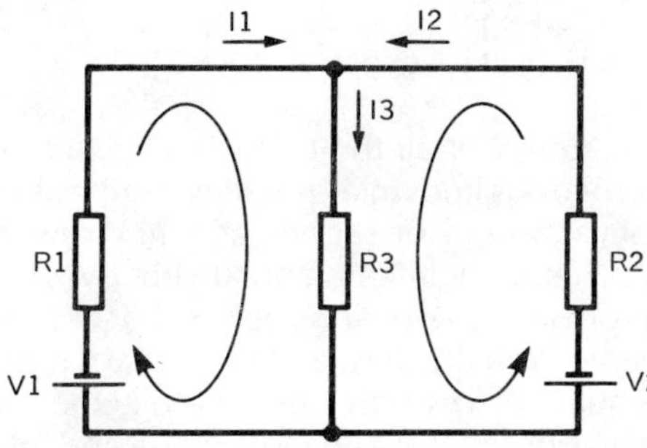

Figure 8.5 A simple network

In the case of the circuit shown in Fig. 8.5, Kirchhoff's law gives

$$I1 + I2 + I3 = 0$$

and the voltage law gives

$$E1 = R1I1 + R3(I1 + I2)$$

and

$$E2 = R2I2 + R3(I1 + I2)$$

Normally we know the values of E1, E2, R1 and R2 and need to know the resulting currents flowing in the three resistors, i.e. I1, I2 and I3. If R1 = 500 ohms, R2 = 200 ohms, R3 = 300 ohms, E1 = 6 V and E2 = 3V we can rewrite the equations as

$$\begin{array}{llll} I1 & + \quad I2 & + \, I3 & = 0 \\ 800I1 & +300I2 & +0I3 & = 6 \\ 300I1 & +500I2 & +0I3 & = 3 \end{array}$$

To find the currents we have to solve this set of simultaneous equations. The usual method is to subtract multiples of rows from one another to reduce the equations to a form that has coefficients of 1 in the diagonal and zeros below. For example, subtracting 800 times the first equation from the second equation gives

$$
\begin{aligned}
I1 \;+\; & I2 \;+\; I3 = 0\\
& -500I2 \;-800I3 = 6\\
300I1 \;+\; & 500I2 \;+\; 0I3 = 3
\end{aligned}
$$

Subtracting 300 times the first row from the third gives

$$
\begin{aligned}
I1 \;+\; & I2 \;+\; I3 = 0\\
& -500I2 \;-800I3 = 6\\
& 200I2 \;-300I3 = 3
\end{aligned}
$$

We have now got rid of all the terms in I1 apart from in the first equation. This process is known as pivoting, and in this case we have removed the first variable in the second and third rows by pivoting on the first row. In general, by pivoting on the Ith row you can get rid of the Ith variable from all the equations below. For example, to remove the second variable from the third equation all we have to do is pivot on the second equation. The only trouble is that the coefficient of the second variable in our second equation isn't 1 and this makes working out what to multiply it by more difficult. The problem is easily solved by multiplying the second equation by 1/−500, giving

$$
\begin{aligned}
I1 \;+\; & I2 \;+\; I3 = 0\\
& I2 \;+1.6I3 = -0.012\\
& 200I2 \;-300I3 = 3
\end{aligned}
$$

Now subtracting 200 times the second equation from the third gives

$$
\begin{aligned}
I1 \;+\; & I2 \;+\; I3 = 0\\
& I2 \;+1.6I3 = -0.012\\
& -620I3 = 5.4
\end{aligned}
$$

Now the solution to the equations can be obtained very easily. From the last row it is obvious that I3 = −5.4/620 or around −9 mA. Using this result in the second equation gives

$$I2 = -0.012 - 1.6I3$$

or around 2 mA. Finally, using both results in the first equation gives a solution of around 7 mA. This process of working back up the equations using results already derived is known as 'back substitution'.

Thus, the solution of any set of simultaneous equations can be seen as pivoting on each row in turn and then using back substitution. Writing a program to perform these operations is fairly simple but first it is worth summarising what pivoting on row I entails:

1. Divide each coefficient in the Ith row by the Ith coefficient in this row. (This reduces the Ith coefficient to 1.)

2. Subtract from the I + 1 row the new Ith row multiplied by the Ith coefficient in the I + 1 row.
3. Repeat step 2 for the I + 2 row, the I + 3 row, etc. until you reach the last row, when the pivot is complete.

Repeating this pivoting operation for I = 1 to N, where N is the number of equations, is all you have to do to reduce the solution to back substitution.

```
10 PRINT "SOLUTION OF SIMULTANEOUS EQUATIONS"
20 PRINT "NUMBER OF EQUATIONS ";
30 INPUT N
40 DIM C(N,N+1)
50 GOSUB 1000
60 GOSUB 2000
70 GOSUB 4000
80 GOSUB 5000
90 GOSUB 6000
100 STOP

1000 FOR I=1 TO N
1010 PRINT "CO-EFFICIENTS FOR EQUATION ";I
1020 FOR J=1 TO N
1030 INPUT C(I,J)
1040 NEXT J
1050 PRINT "CONSTANT =";
1060 INPUT C(I,N+1)
1070 NEXT I
1080 PRINT
1090 RETURN

2000 FOR I=1 TO N-1
2010 GOSUB 3000
2020 NEXT I
2030 C(N,N+1)=C(N,N+1)/C(N,N)
2040 C(N,N)=1
2050 RETURN

3000 C=C(I,I)
3010 FOR J=1 TO N+1
3020 C(I,J)=C(I,J)/C
3030 NEXT J
3040 FOR J=I+1 TO N
3050 C=C(J,I)
3060 FOR K=1 TO N+1
3070 C(J,K)=C(J,K)-C(I,K)*C
3080 NEXT K
3090 NEXT J
3100 RETURN

4000 FOR I=1 TO N
4010 FOR J=1 TO N+1
4020 PRINT TAB(J*10);C(I,J);
```

```
4030 NEXT J
4040 PRINT
4050 NEXT I
4060 RETURN

5000 DIM A(N)
5010 A(N)=C(N,N+1)
5020 FOR I=N-1 TO 1 STEP -1
5025 A(I)=C(I,N+1)
5030 FOR J=I+1 TO N
5040 A(I)=A(I)-A(J)*C(I,J)
5050 NEXT J
5060 NEXT I
5070 RETURN

6000 FOR I=1 TO N
6010 PRINT "VARIABLE ";I;" = ";A(I)
6020 NEXT I
6030 RETURN
```

The program given above will solve any number of simultaneous equations using pivoting. The main program consists of the usual list of subroutine calls, apart from lines 10 to 40 which set up an N by N + 1 array C which is used to hold the coefficients (N is the number of equations). Subroutine 1000 reads in the coefficients of each equation in turn and stores them in C. No opportunity is given to the user to correct mistakes, so type carefully! Subroutine 2000 carries out the pivoting operation on each of the rows in turn. The FOR loop at lines 2000 to 2020 carries out a pivoting operation on the Ith row, for row 1 to N − 1, by calling subroutine 3000. Lines 2030 and 2040 simply reduce the work involved in back substitution by changing the final equation's coefficient to 1 after all the other pivoting operations have been completed. Subroutine 3000 carries out a pivot on the Ith row and as already mentioned is used repeatedly by subroutine 2000. Lines 3000 to 3030 changes the Ith row so that C(I,I) is 1. That is, it carries out step 1 of the pivoting.

The rest of the subroutine is in the form of a nested pair of FOR loops. The inner loop, lines 3060 to 3080, subtracts the correct multiple of the Ith row from the Jth row, where J varies from I + 1 to N, as determined by the outer loop, lines 3040 to 3090. When this subroutine is finished the Ith variable has been eliminated from all the equations I + 1 to N.

Subroutine 4000 prints the final version of the equations ready for back substitution and subroutine 5000 actually carries out this back substitution. Finally, subroutine 6000 prints the answers that are stored in the array A.

That is all there is to solving simultaneous equations by computer. Notice that no details of the back substitution subroutine have been given. The reason for this is that the theme of this section is that anything that you know how to work out by hand you should be able to turn into a computer program. With this in mind you should find subroutine 5000 easy to work out for yourself. If you have any difficulty, study the pattern of what you multiply by what when you do the back substitution by hand.

If you type in the coefficients of the simultaneous equations given earlier you will find that the exact solutions are printed out without any extra effort on your part. However, if you try a set of equations such as

$$
\begin{aligned}
I1 &+ 0I2 + 3I3 = 0 \\
4I1 &+ 0I2 + 4I3 = 12 \\
I1 &- 3I3 - I3 = 6
\end{aligned}
$$

you will find that the program fails to give a solution and instead crashes at line 3020. The reason for this is that when you try to pivot on the second equation, the second coefficient is zero and dividing by zero is not defined. The solution is simply to rearrange the order of the equations so that the second equation becomes the third. This sort of problem can crop up at any time during the solution of simultaneous equations using pivoting.

Another related problem is that the accuracy of the method depends on the size of the coefficient that you are dividing by — the larger the more accurate the result, and vice versa. The obvious solution to both these problems is to rearrange the array so as to produce the largest element in C(I,I), just before the pivoting on the Ith row. To add this operation to the program is easy because of the way that it has been written using subroutines. The only change to the existing program is to add line 2005

```
2005 GOSUB 7000
```

to subroutine 2000. The new subroutine starting at line 7000 (see page 92) checks down the elements in the Ith column below the current C(I,I). If it finds one bigger it exchanges the two rows. With this addition the program will solve any simultaneous equations that have solutions.

Large programs

The last two programs in this chapter, the plotter and the simultaneous equation solver, are normally considered to be quite large and

```
7000 M=C(I,I)
7010 K=I
7020 FOR J=I TO N
7030 IF ABS(M)>ABS(C(J,I)) THEN GOTO 7040
7035 K=J
7036 M=C(J,I)
7040 NEXT J
7050 IF K=I THEN RETURN
7060 FOR J=1 TO N+1
7070 T=C(I,J)
7080 C(I,J)=C(K,J)
7090 C(K,J)=T
7100 NEXT J
7110 RETURN
```

difficult programs. However, if you split a large program down into subroutines then it can be tackled as a series of small programs. Looked at in this way, your large programs offer no new difficulties and you should be able to go on to write useful programs of your own. However, it must always be kept in mind that the computer cannot do a calculation, or anything else, that you cannot do — at least in principle. As with the simultaneous equation example it is often helpful to work out an example by hand (and calculator) to make sure that you understand the method. This also has the advantage that you have a test calculation for your program, for which you already know the answer.

The best advice is always to break what you are doing down into small steps before you begin to program it. You could almost say that the programmer's motto is 'divide and conquer'!

9

Advanced BASIC

Don't be alarmed by the title of this chapter. In this instance, advanced doesn't mean difficult. To be more accurate, this chapter is about non-standard features of BASIC. So far we have used only those parts of BASIC that you can normally rely on finding on any machine. The value of staying within this subset of BASIC is that any program that you write should be easy to transfer to another machine. There are reasons, however, why you should know about the sorts of extra features that you can find on some machines. Sometimes the only way you can make a program fit into a particular machine is to use some special statement. It is also a good idea to know what the limitations of your own machine are by finding out what your BASIC lacks.

Prompts

If you look back at some of the earlier example programs you'll notice that INPUT statements are very often preceded by PRINT statements. You normally print out a question and then wait for the answer. To save space some versions of BASIC allow a literal string to be included in an INPUT. For example:

```
10 INPUT "HOW OLD ARE YOU",A
```

will print HOW OLD ARE YOU? and then wait for the answer to be typed in. The literal string is called a prompt. The trouble with using this very convenient form of INPUT is that even the BASICs that do allow it can't agree on its exact form. Some use a semicolon instead of a comma to separate the prompt from the input variable, some print a question mark and some don't.

Print formatting

We have seen that the TAB function can be used in combination with a semicolon and a comma to control where things are printed on the screen. So far we haven't discovered any way to control *how* things are to be printed on the screen. For example, if we use, say, "PRINT A", the result displayed might be .3 or .3333333333 or 0.3333333, etc. We have no control over how many decimal places will be printed or how many leading zeros will be used.

This problem is overcome in Microsoft BASIC by the PRINT USING statement, the format of which is:

```
PRINT USING "format string"; print list
```

The format string may be either a literal string or a string variable and defines how the variables in the print list will be printed. The range of formats allowed is very machine-dependent and very wide. The most that can be done here is to give a simple example.

You can 'draw' a picture of the number using # to represent each digit; e.g. ###.### is a picture of a six-digit number with three digits after the decimal point. So

```
10 PRINT USING "###.###";.78,200.45,.3333333
```

prints

```
0.780   200.450   0.333
```

If the number has too few digits to fill the picture then blanks are added from the right. If a digit is specified before the decimal point then that digit will always be printed even if it is zero.

The PRINT USING command is very powerful and provides a way of producing very neat screenfuls of numbers. For technical programming, however, the most common requirement is to print a number to a given number of decimal places and you can do this without PRINT USING. Instead, the following can be used:

```
10 DEF FNA(X,N) = INT(X*N)/N
20 PRINT FNA(3.3333,1000)
```

The function FNA(X,N) will print X to a number of decimal places equal to the number of zeros in N. (For this to work N must be 10,100,1000,100000, etc.) If you don't have a function definition statement in your BASIC then simply use INT(X*1000)/1000 every time you need it.

ELSE

The BASIC IF statement has one extra trick waiting to be used. Some versions of BASIC allow an extended form of IF:

```
IF condition THEN "BASIC statement" ELSE "BASIC statement"
```

This is a much more symmetrical form of the IF statement. If the condition is true, then the BASIC statement following THEN is carried out. If the condition is false, then the BASIC statement following ELSE is carried out. That is, only one of the two statements (the one after THEN or the one after ELSE) is executed. Thereafter, control passes to the next statement (usually the next program line). For example:

```
10 INPUT A
20 IF A = 0 THEN PRINT "A IS ZERO" ELSE PRINT "A IS NOT ZERO"
30 GOTO 10
```

is the same as

```
10 INPUT A
20 IF A = 0 THEN PRINT "A IS ZERO"
30 IF A<>0 THEN PRINT "A IS NOT ZERO"
40 GOTO 10
```

As you can see, the IF–THEN–ELSE form is easier and neater but not essential.

While we are on the subject of IF statements and tests it is worth considering the problems of accuracy in BASIC.

Accuracy

Try the following seemingly reasonable program:

```
10 I = 1
20 I = I – 1/3
30 IF I = 0 THEN STOP
40 GOTO 20
```

The meaning of the program is simple: set I equal to 1 and subtract 1/3 until you have nothing left, i.e. until you reach zero. If you follow the program through with pencil and paper you should have a value of 2/3 for I the first time you reach line 30, 1/3 on the second and 0 on the third. So the program should stop after three times around. If you've run the program the chances are that you are puzzled because the program went round and round forever. What went wrong?

If you insert 25 PRINT I you should be able to see what is going on. The trouble is that 1/3 isn't stored accurately inside your machine, it's probably something like .33333333 which is *less* than 1/3. Because of this, after three times through the loop I is *close* to zero but it's not *exactly* zero. This sort of problem can be overcome by using the golden rule: never test for *equality* between numbers that result from a calculation. If, however, only whole numbers are used throughout (i.e. no fractions, etc.) then the problem does not arise.

The solution to the problem of inaccuracy in fractional numbers is to change any test for equality into a test for 'closeness'. If two numbers are different by an amount that is about the same size as the accuracy of your machine's calculations then they are as good as equal. For example, try:

```
10 I = 1
20 I = I – 1/3
30 IF I<.000001 THEN STOP
40 GOTO 20
```

This will work on your machine if the accuracy is better than 0.000001. If not, then adjust the number of zeros following the decimal point until the program works. In general, you can change

```
10 IF A = B THEN . . .
```

into

```
10 IF ABS(A – B)<.000001 THEN . . .
```

Multi-statement lines

A great many versions of BASIC will allow you to write more than one statement on a line, each one separated by a colon. Thus

```
10 A = 1:B = 2:IF A = 0 THEN GOTO 50
```

has the same effect as

```
10 A = 1
20 B = 2
30 IF A = 0 THEN GOTO 50
```

This use of the colon extends the default flow of control from just 'down the page' to left to right and then down the page. This is a useful feature as long as you keep in mind the following two points. First, long BASIC lines can be difficult to read and, because error messages generally supply a line number, it can also be difficult to tell which part of a long line is causing trouble. And second, most versions of BASIC place a maximum size on the length of the longest line that you can type in (usually 80 characters) and this in turn limits the longest multiple statement line that you can use.

Before using multiple statement lines it is worth finding out a little more accurately where GOSUBs return to. For example, on your computer what does the following program print?

```
10 A = 0:GOSUB 1000:A = 1:GOSUB 1000
20 A = 2:GOSUB 1000
30 STOP

1000 PRINT A
1010 RETURN
```

If the first GOSUB returns to the next line (i.e. line 20) then you will see 0,2 printed on the screen. However if the first GOSUB returns to complete the rest of the multi-statement line then you will see 0,1,2 printed on the screen.

One application of the multi-statement line is as part of the IF–THEN–ELSE statement. If you have two lists, one of which you want to carry out if the condition is true and the other that you want to carry out if the condition is false, then you could use

```
IF condition THEN list 1 ELSE list 2
```

where list 1 and list 2 are BASIC statements separated by colons. However, if the lists are at all long then not only does the IF become difficult to read and debug, but also there is not much chance that BASIC will allow you to enter it! A better solution to this problem is to group the lists together as two BASIC subroutines and use

```
IF condition THEN GOSUB list 1 ELSE GOSUB list 2
```

where in this case list 1 and list 2 are the line numbers of the first statement in each list.

DATA

The DATA statement can be considered to be fairly standard BASIC — you'll find it in Microsoft BASIC and it was in the first-ever version of BASIC (Dartmouth BASIC). The only important machine that doesn't have a DATA statement is the ZX81. This, coupled with the fact that you can do without it, is the main reason for ignoring the DATA statement until now.

The DATA statement is a way of initialising variables to values other than zero or the null string. The form of the DATA statement is

DATA list of constants

The list of constants is simply a set of values separated by commas and can be whole numbers, fractions or even strings. For example:

```
10 DATA 1,2.3,"HELLO",-5.9
```

Notice that the string constant must be surrounded by quotes.

To make any use of the DATA statement we have to assign the constants to variables. This is done using the READ statement. The form of the READ statement is

READ list of variables

The list of variables is simply a list of valid variable names separated by commas. When a READ statement is executed the program is searched for a DATA statement. If none is found then an error is reported. If one is found then the first constant in the constant list is assigned to the first variable in the variable list, the second constant to the second variable, and so on until all the variables in the READ statement have been used up. For example, if the following READ statement is in the same program as the previous example of a DATA statement

```
20 READ A,B,A$,C
```

then A would be 1, B would be 2.3, A$ would be HELLO and C would be −5.9.

A few things have to be remembered when using DATA and READ statements. First, there must be at least as many constants as variables. You can have constants that you never use, but you must have a constant for each variable. If the constant list runs out before the variable list then BASIC looks for another DATA statement and if one cannot be found an error is reported. Second, the type of each constant must match the type of the variable it is to be assigned to, otherwise an error is reported. In other words, a numeric variable needs a numeric constant and a string variable a string constant. Finally, if a second or subsequent READ statement is encountered later in the program, the assignment of constants to variables continues from where the last READ statement left off . . . i.e. you do not go back to the start of the constant list for every READ statement.

This last observation raises the question of how you could go back to the start of a list of constants. The answer is the RESTORE statement. If you simply issue the command RESTORE the next READ statement will start its assignment from the first constant in the first DATA statement in the program.

The use of DATA and READ statements can save quite a bit of program space if used to initialise variables rather than a list of assignments. To see that this is true go back to the morse code program given in Chapter 4 and change it to use DATA and READ statements to initialise the array C$.

Specialised commands

There are lots of other BASIC statements that you can find in your machine's manual but nearly all of them are specific to your machine. I'm not suggesting that they are not useful — just specialised. Most of the time you can get by using just the set of BASIC statements that we have discussed. However, the one thing that you should not do is to suppose that you *know* BASIC — you must always check to find out what special peculiarities your machine has.

In particular, if you use either a tape or disk to store data you will have to find out how your BASIC handles data files and this is one of the most machine-dependent features of BASIC. Similarly, if you want to make the best use of your graphics or screen handling, you will have to look up instructions like PLOT or POS. After reaching this point you should have no trouble with understanding your manual. If you do, then remember that computing is an experimental science — guess how it works and then try it to see!

How BASIC works — interpreters and compilers

In Chapter 1 it was explained that BASIC is a high level language that has to be converted into a low level language or machine code before the computer can obey its instructions. There are in fact two different ways of carrying out this conversion — BASIC can be 'interpreted' or it can be 'compiled'. Far from being a matter of academic importance, the way that a computer language is implemented has a great deal of effect on the performance and ease of use of the language.

The easier of the two methods to understand is the compiler. A compiler takes the program that you have written and translates it to machine code. Only after it has been completely translated is your program run. Thus, using a compiler has a number of distinct steps. First you have to create the text of the program that the compiler will translate. This usually involves using another program called an 'editor' to type in and correct the lines of BASIC. Once you have the text then the compiler can translate this to machine code, which can then be run. These steps can make program development and debugging take a great deal of time. However, there is one great advantage in using a compiler. As your program is completely converted to machine code it runs very fast and once you have a working program there is no need to re-compile it every time you use it.

If you use a microcomputer with Microsoft or one of the other versions of BASIC that have been mentioned throughout this book then these stages of editing, compiling and running will not sound familiar. When you use BASIC you can type in lines, edit old lines and then type RUN and try your program out without any switching between editors and compilers. The reason for this is that the most popular method for implementing BASIC is to use an interpreter. An interpreter makes no attempt to convert your entire program into machine code. Instead what it does is to build up the text of your program in the computer's memory and when you type RUN it examines it line by line to find out what you want it to do. Although no conversion to machine code actually takes place, in many ways you can think of this interpreting of each line as a line-by-line translation of your program as opposed to what a compiler does when it translates the entire program before letting you run it.

To clarify the difference, consider the way a compiler and an interpreter would implement the very simple BASIC statement GOTO 1000. The compiler would translate all of the program to machine code and so, in the final form of the program, the BASIC that once made up line 1000 would be a collection of machine code instructions that resided at a particular address in the computer's memory. The command GOTO 1000 would be translated into the equivalent machine code instruction to transfer control to the machine code that line 1000 generated. When the computer encountered this transfer of control instruction while running the program it would obey it immediately. An interpreter, however, wouldn't change any part of the program to machine code. Instead, when you type RUN an interpreter examines your program line by line and obeys it. When it reaches the line GOTO 1000 it first examines the word GOTO and tries to find it in a list of BASIC instructions that it keeps in memory. When it discovers that it is a GOTO instruction it then examines the characters that follow the word GOTO and assembles them into a line number which it then searches the rest of the program for. In a large program this search itself can take quite some time, because a considerable number of lines may have to be examined before line 1000 is found.

You should be able to see that interpreters and compilers differ in the maximum speed that you can achieve. What is less obvious is the ease that an interpreter brings to writing and testing a program. The ideal situation is to use an interpreter to develop a program and a compiler to translate it to machine code for use. This ideal of an interpreter/compiler pair is available for many machines, including all those running CP/M, but you will find that in general compilers are expensive.

Where next?

The topic of advanced BASIC leads us on to consider what language you should learn next. There is no denying that BASIC has many faults — not quite as many as its arch-critics would have us believe but enough to make another language a proposition worth considering. It is worth pointing out that much of the criticism of BASIC is either founded in the misunderstanding that a particularly limited or early version of BASIC is *the* BASIC that is still current, or is based on a misguided confusion of good programming with good programming languages. You can write good programs in any language and, while it is true that some languages encourage good programming by providing better facilities, it is difficult to find a language that is at the same time academically respectable and technically useful.

One of BASIC's biggest competitors is the academically acclaimed Pascal. It has better subroutines and functions than BASIC and its control statements are designed to encourage structured programming (see Chapter 6). However, it lacks any 'raise to the power' operator or string handling functions and it is not generally available in a form that takes advantage of any special hardware that your computer might have — graphics and sound being two of the most attractive.

Instead of moving over to a completely different language such as Pascal you might be tempted by languages such as COMAL which are mixtures of Pascal and BASIC. These are indeed worth considering but there are dialects of BASIC, BBC BASIC for example, that are almost as advanced but still called BASIC and much more widely used.

For technical applications it is worth looking back to an old and well-tried language — FORTRAN. Although this language has been in use since the early days of computers it is still popular for technical applications. It has a good subroutine and function facility and supports not only complex numbers but also complex arithmetic.

Later versions of FORTRAN (FORTRAN 77, for example) are very close to BASIC, in spirit at least, and you should have no trouble using both.

The last paragraph should not be taken to mean that I advocate ancient languages! The fact of the matter is that it is high time that a new computer language gained some of the ground that BASIC currently occupies as the most popular language. However, at the time of writing it is difficult to see a single clear winner in terms of simplicity of use, technical facilities offered, and the availability of standard implementations. The newest computer language ADA may be waiting in the wings to sweep FORTRAN and BASIC into a museum — but I doubt it.

While there is safety in numbers, BASIC is best.

Self-test questions

1 Write a routine using input prompting to ask the user for their name and the date.

2 Write a program to print a list of random numbers to three decimal places.

3 Write a program to discover the accuracy of your computer.

4 Go through your computer's manual looking for extra commands and features of its BASIC and consider how their use might improve or extend your programs.

10

The computer in electronics

In this final chapter we look at the many ways that the computer can be used in electronics. In any overview of this sort there is always the problem of seeing things from the wrong point of view. In this case it would be all too easy just to list the large and costly computer-based equipment available commercially and ignore the smaller innovations using low cost personal computers.

The contribution of the amateur

Electronics is a very popular hobby — although it's not as popular as it used to be. The reason for this is that the microcomputer is taking over the place that electronics once held as the prime technical hobby. There has always been an interest in computers inherent in electronics. Even back in the dark days of transistors, electronics magazines published designs for simple computers. Only the limits on costs and size held things back. As soon as the microprocessor appeared on the scene it was used by electronics hobbyists to produce microcomputers. The early microprocessors were very simple devices and needed a lot of work to turn them into useful computing machines — but this effort was the source of the fun that combines electronics and computing. The first home computer — the Altair — started as a constructional project in an American electronics magazine and resulted in the vast range of machines on sale today. Because of this vast range there is now no longer any financial reason to build your own machine — you can buy a commercial model for not much more than the cost of the components. It is true that some people are still interested in building their own computer just for the challenge and satisfaction. However, just because you buy equipment off the shelf it doesn't mean that you're not interested in electronics.

In the world of amateur radio there is the same split — some build and some buy equipment, but both groups are united by the way they *use* the equipment. This unity is not equally true of computing/electronics enthusiasts. Some electronics amateurs discover that once they have acquired a computer their interest in electronics declines. The software side of computers becomes so absorbing that any ideas of circuit construction are replaced by program design. After all, software has many advantages over hardware. It requires no special equipment (over and above the computer), it is easy and safe to experiment with, it is easy to change and once you've worked it out then mass-producing it for other people to use is relatively cheap. Compare this with the problems of hardware construction — soldering irons, faulty components, burned out components, etc. — and you'll see that it's not surprising that once the intricacies of a program language have been mastered some people never look back.

The other side of the story is that some electronics enthusiasts discover that software is interesting but not as interesting as electronics. There are a number of distinct ways that this renewed interest in electronics can go. If an interest in computer hardware still remains, then the task of improving the off-the-shelf equipment might occupy some time. A more rewarding task is to connect other pieces of equipment to the computer — and this takes the hobbyist into the difficult area of interfacing. There is a definite lack of skilled people able to tackle both the hardware and the software side of computing and this is what is required to solve an interfacing problem.

It is however possible that, after satisfying a curiosity about computers by buying one, any interest in computer hardware vanishes and an older interest in some other area of electronics returns. In the same way, the professional may see the computer more as a tool than an end in its own right. The question is, then, what can the computer do for other areas of electronics?

Interfacing

One of the apparently most difficult areas of computer use is the connection of personal computers to other pieces of equipment — generally referred to as interfacing. The trouble with interfacing is that it is such a wide subject. You need to know a great deal about computers and electronics to be able to see the possibility of connecting a computer to something, let alone do it! The things that people have done in the past range from the legendary central heating control to simple robotics. With an interfacing project there are normally three areas of difficulty. Obviously, you have to design and test the electronics and the software but there is also the problem of the interaction between the two. That is, the hardware may work perfectly and the software may be well written and free from bugs, but the way the software uses the hardware may be incorrect. This interaction is the main reason why interfacing is difficult. Imagine the situation: you have just finished wiring up your computer-controlled disco lights and everything is working well except that the red light flashes when the blue one should. Is the fault in the hardware or the software?

You can become involved in computer interfacing projects with the simplest and cheapest of home computers. The ZX81, for example, has a range of add-ons that simplify interfacing and I've heard of it being used for all sorts of things, including a model railway controller, a burglar alarm, and a disco sound and light synchroniser. If you want to go in for interfacing type projects then, apart from your computer, you will need a reasonable knowledge of digital electronics, including how A-to-Ds and D-to-As work, a good understanding of programming, possibly down to the assembly language level, and you will need to know how your computer works!

ECAP

It would speed up the design and construction of any electronic circuit if it were possible to describe it to a computer and have the computer describe back to the user how the circuit works. This is possible through the use of ECAP. ECAP stands for Electronic Circuit Analysis Program. To be able to do this depends entirely on having the necessary software. There are two ways of acquiring this software: you can buy it or you can write it. If you decide to write it then of course you run the risk of spending so long developing the software that you never get round to constructing any circuits. The trouble is that buying it isn't really practical either because there isn't any good cheap ECAP software for micros.

All in all, complete ECAP facilities are a fairly ambitious request for a small home computer to fulfil. If you do decide to try to write your own general ECAP software then there are two main problems that you will have to overcome. The first is the description of your circuit to the computer. You could adopt the traditional approach of asking for a list of components and interconnections but a more ambitious and satisfactory method would be to use a light pen to draw the circuit directly on the screen. Either of these methods require a fair bit of programming to achieve results worth having.

The second main problem with general ECAP is actually working out the voltages and currents at the various points in the circuit. This is difficult even if the only components used in the circuits are resistors.

The trouble is that to work out the current through any component you need to know the voltage across it. Normally you only specify the total voltage across the circuit, i.e. the supply voltage, and this distributes itself across the components in the circuit. The way it spreads itself throughout the circuit depends on the values of the components and the way they are connected together. To calculate these values you have to write down the basic equations that the circuit must obey and then solve them.

For circuits involving resistors and voltage sources, the equations are always simple simultaneous equations of the type discussed in Chapter 8. Of course any general ECAP software would have to deal with components other than resistors and things then start to get very tricky because the resulting equations are difficult to solve even with a computer.

Even though the technical difficulties of ECAP are challenging they are at least fairly well understood. In fact all we are doing is getting the computer to apply the traditional methods of analysing linear circuits that can be found in any textbook. A much more difficult problem lies in trying to provide a good user interface to such a package. This is something that we cannot look up in textbooks. There is a second ECAP area that is much more difficult to approach using any general method — that of non-linear circuits. The analysis of non-linear circuits using computers is very important because often there is no way that the circuit could be analysed without a computer.

Special circuits

Although the general ECAP problem is difficult you can produce programs to help with the design of standard circuits. For example, the program given in Chapter 3 for calculating the component values for the 555 timer is far from the general ECAP problem, but if you want to design a timer circuit then a simple program is just as good. Likewise the Zener diode design program in Chapter 8 is more useful in practice than a comprehensive ECAP program if you are in the business of designing power supplies.

You could add to the list of standard circuit design programs until the general program was almost unnecessary. All you have to do to write a standard design program is to get hold of the data sheet for the IC/transistor/diode in question and transfer the calculations that you would normally do by hand to the program. The only temptation to be resisted is giving far too much information. Do not work out circuit parameters that are never likely to be needed! However the computer does give you the opportunity to apply the standard design equations in ways that hand calculation would have ruled out. For example, by repeating the calculations for a range of values you can supply information about the stability of the circuit without using calculus.

Constructional aids

The area of using a computer as an aid to construction is probably the least explored and the most promising. At the very simple level, you can do things like decoding resistor/capacitor colour codes and working out transformer windings, etc. There are more interesting applications, however.

The job of producing pin-to-pin wiring tables for wire wrapping or soldering is a time-consuming and tedious task. In fact it's so tedious that most amateurs and prototype builders avoid it all together by wiring circuits directly from a circuit diagram. If you have a program that you can use to record the number of components in the circuit, the number of pins each has and where each pin connects, then you can let the program check that no silly connections, such as a pin connected to itself, have been specified and then print out a check list of connections and the order in which they should be done. You could even use the machine to record each connection as and when you make it.

Other areas that the computer can help with are a little more difficult and costly to realise and so are confined to large production units. In the electronics industry large computers are used for circuit layout and testing. The appeal of both these applications is obvious. Imagine sitting in front of a small computer, light pen in hand, with a printed circuit board layout displayed on the screen. You point to two pins and they are instantly (well, almost instantly) connected together. When you have finished, your printer produces a plan of the printed circuit board, a layout plan, a wiring table, a circuit diagram . . . The problems involved in realising such a piece of software for a low-cost home computer are large but many would be satisfied with something that approached the ideal.

The second area of possible computer help — fault finding — is also a difficult one to reduce in scale and the local service centre still tends to use the traditional instruments, meter and scope, to find faults. In order even to start to automate this tedious business you need to couple a small computer to an A to D converter and write a program that tests the equipment stage by stage. To do this for a general piece of equipment is difficult but if you want to service a particular unit then the problem is not so bad. At a slightly less ambitious level you could write a program that guides the servicing rather than carries it out. The program would ask questions like 'Have you checked the fuse?' and would accept your answers, trying all the time to decide what was wrong. There are developments going on in the area of artificial intelligence that promise to remove all of the skill in servicing to a computer program but we will have to wait to see if this has any practical impact.

On a more mundane level, small computers are beginning to change the range and cost of measuring instruments available. At first microprocessors worked their way into measuring equipment as dedicated controllers that the user would know nothing about. Now there is a growing trend for the existence of the micro to be acknowledged and the user encouraged to make use of it. For example, you can now buy an add-on card for the Apple II that turns it into a two-channel storage scope (10 MHz) for a price that you would expect to pay for a standard scope. Computer-based instruments are not only cheaper, they are better — in the sense that they can be used in ways that traditional equipment could never be. For example, a computer-based scope can be set to trigger on a complex wave form, to average periodic signals, to lock on to non-periodic signals, and so on.

Electronics as software

So far we have considered the role of the computer as a tool in realising new pieces of electronic equipment. However, there is the point of view that the computer itself is the best way of actually realising any new hardware. In this sense it is the raw material for all electronics!

As an example consider the problem of designing and constructing a new radio receiver. How this would be achieved using standard components is something that we all know about; but what about the pure computer solution? If you imagine a very fast computer with an A to D converter connected to an aerial and a D to A converter connected to a speaker then this is all the hardware we need for the new radio. The rest of the 'construction' would be carried out entirely in software. The digitised signal from the aerial would be processed by digital filtering. It would be amplified and the audio signal recovered using nothing but arithmetic. Then the stream of numbers representing the audio signal would be fed to the D to A converter and out into the external world via the speaker. The advantages of this approach are obvious: any changes to the design are simply software changes, the imperfections introduced by traditional components are avoided and the same computer chip could be used for many different products, so lowering the overall price.

The reason why this approach isn't used at the moment is equally obvious: there are no computer chips that work fast enough to make the scheme possible! This may always be the problem in that, no matter how fast computers become, the components that make them up will always work faster and this in turn implies that direct implementation by electronics will always offer a speed advantage. However, for some applications this speed advantage will be irrelevant and the computer will slowly become the standard method of producing new designs. Already the digital processing of audio signals is as important as analog methods, and the computer is by far the best way of implementing any complex digital logic unless the highest speed is required.

As time passes the role of the computer *in* electronics will diminish for the simple reason that the computer will *become* electronics!

Answers to self-test questions

Chapter 2

1 (a) 17 (b) 2.5 (c) 8.4 (d) 2.5 (e) 0.25

2
```
10 PRINT "LENGTH OF SIDE =";
20 INPUT L
30 A = L*L
40 PRINT "AREA =";A
```

3
```
10 PRINT "LENGTH OF FIRST SIDE =";
20 INPUT L
24 PRINT "LENGTH OF SECOND SIDE =";
26 INPUT S
30 A = L*S
40 PRINT "AREA = ";A
```

4
```
10 PRINT "INDUCTANCE =";
20 INPUT L
30 PRINT "WIRE DIAMETER =";
40 INPUT D
50 PRINT "RADIUS OF CORE =";
60 INPUT R
70 N = (0.4*L*D + 0.16*L*L*D*D + 0.0033*R*R*R*L)/(0.0031*R*R)
80 PRINT "NUMBER OF TURNS = ";N
```

Chapter 3

1
```
10 FOR I = 1 TO 5 STEP 0.1
20 M = I*25
30 PRINT I;" INCHES = ";M;" MILLIMETRES"
40 NEXT I
```

2
```
 1 PRINT "STARTING VALUE =";
 2 INPUT S
 3 PRINT "FINAL VALUE =";
 4 INPUT F
 5 PRINT "STEP SIZE =";
 6 INPUT C
10 FOR I = S TO F STEP C
20 M = I*25
30 PRINT I;" INCHES = ";M;" MILLIMETRES"
40 NEXT I
```

3
```
 10 PRINT "RESISTANCE AT ZERO C =";
 20 INPUT R
 30 PRINT "TEMPERATURE COEFFICIENT =";
 40 INPUT A
 50 PRINT "STARTING TEMPERATURE =";
 60 INPUT S
 70 PRINT "FINAL TEMPERATURE =";
 80 INPUT F
 90 FOR T = S TO F
100 M = R*(1 + A*T)
110 PRINT "AT ";T;"C THE RESISTANCE = ";M
120 NEXT T
```

4
```
 10 INPUT R
 20 IF R>=1000 THEN GOTO 50
 30 PRINT R;"OHMS"
 40 GOTO 10
 50 IF R>=1000000 THEN GOTO 90
 60 R = R/1000
 70 PRINT R;"K OHMS"
 80 GOTO 10
 90 R = R/1000000
100 PRINT R;"M OHMS"
```

5
```
110 GOTO 10
180 FOR J = S TO F STEP (F−S)/N
190 PRINT "FOR R = ";J;"K OHMS C IS ";T/(1.1*J);"MICRO FARADS"
200 NEXT J
```

Chapter 4

1
```
10 DIM A(10)
20 FOR I = 1 TO 10
30 INPUT A(I)
40 NEXT I
50 FOR I = 10 TO 1 STEP −1
```

```
60 PRINT A(I)
70 NEXT I
```

2
```
 4 PRINT "HOW MANY ";
 6 INPUT N
10 DIM A(N)
20 FOR I = 1 TO N
30 INPUT A(I)
40 NEXT I
50 FOR I = N TO 1 STEP −1
60 PRINT A(I)
70 NEXT I
```

3
```
10 INPUT A$
20 FOR I = LEN(A$) TO 1 STEP −1
30 B$ = MID$(A$,I,1)
40 PRINT B$;
50 NEXT I
```

or in ZX BASIC:

```
10 INPUT A$
20 FOR I = LEN(A$) TO 1 STEP −1
30 LET B$ = A$(I,I)
40 PRINT B$;
50 NEXT I
```

4
```
 10 DIM A$(10)
 20 A$(1) = "BLACK"
 30 A$(2) = "BROWN"
 40 A$(3) = "RED"
 50 A$(4) = "ORANGE"
 60 A$(5) = "YELLOW"
 70 A$(6) = "GREEN"
 80 A$(7) = "BLUE"
 90 A$(8) = "VIOLET"
100 A$(9) = "GREY"
110 A$(10) = "WHITE"
120 PRINT "FIRST COLOUR =";
130 INPUT C$
140 C = 0
150 FOR I = 1 TO 10
160 IF C$ = A$(I) THEN C =I
170 NEXT I
180 IF C = 0 THEN GOTO 120
190 R = C − 1
200 PRINT "SECOND COLOUR =";
210 INPUT C$
```

```
220 C = 0
230 FOR I = 1 TO 10
240 IF C$ = A$(I) THEN C = I
250 NEXT I
260 IF C = 0 THEN GOTO 200
270 R = R*10 + C − 1
280 PRINT "THIRD COLOUR =";
290 INPUT C$
300 C = 0
310 FOR I = 1 TO 10
320 IF C$ = A$(I) THEN C = I
330 NEXT I
340 IF C = 0 THEN GOTO 280
350 R = R*10 ↑ (C − 1)
360 PRINT "RESISTANCE =";R;" OHMS"
370 GOTO 120
```

In ZX BASIC change line 10 to DIM A$(10,6), add line 15 DIM B$(6) and, of course, prefix the assignment statements with LET.

Chapter 5

1

```
 10 PRINT "START =";
 20 INPUT S
 30 PRINT "END =";
 40 INPUT E
 50 PRINT "STEP =";
 60 INPUT C
 70 FOR I = S TO E STEP C
 80 A = SQR(I)
 90 PRINT "THE SQUARE ROOT OF ";I;" IS ";A
100 NEXT I
```

2

```
 10 PRINT "START =";
 20 INPUT S
 30 PRINT "END =";
 40 INPUT E
 50 PRINT "STEP =";
 60 INPUT C
 70 FOR I = S TO E STEP C
 80 A = SIN(I)
 85 B = COS(I)
 90 PRINT "ANGLE (IN RADS) = ";I;" SIN = ";A;" COS = ";B
100 NEXT I
```

3

```
10 DEF FNL(X) = LOG(X)*0.434295
```

4
```
10 DEF FNS(X) = X − X↑3/6 + X↑5/120−X↑7/5040
20 INPUT X
30 A = FNS(X)
40 B = SIN(X)
50 PRINT A,B,A − B
60 GOTO 20
```

You will find that A and B are very close as long as X is small. In other words, FNS is one way of calculating SIN(X). Indeed, it may be the way that your computer actually uses!

Chapter 7

1

A B C	A AND B AND C	(A AND B) OR (A AND C)	A AND (B OR C)
0 0 0	0	0	0
0 0 1	0	0	0
0 1 0	0	0	0
0 1 1	0	0	0
1 0 0	0	0	0
1 0 1	0	1	1
1 1 0	0	1	1
1 1 1	1	1	1

2 IF V>=L AND V<=H THEN in range

3
```
10 T = −1
20 F = 0
30 FOR R = T TO F
40 FOR S = T TO F
50 FOR Q = T TO F
60 PRINT R;"  ";S;"  ";Q;"  ";NOT(R AND NOT(Q AND S))
70 NEXT Q
80 NEXT S
90 NEXT R
```

This is the truth table of an RS flip-flop. Notice that true is assumed to be −1 and false 0.

Chapter 9

1
```
10 INPUT "NAME",N$
20 INPUT "WHAT IS YOUR DATE OF BIRTH",D$
30 PRINT N$,D$
```

2
```
10 R = RND(0)
20 PRINT INT(R*1000)/1000
30 GOTO 10
```

(or you could have used PRINT USING)

3
```
10 L = 1
20 S = L + 1
30 IF L = S THEN PRINT L
40 L = L*10
50 S = L + 1
60 GOTO 30
```

The first value of L that is printed is so large that L is not detectably different from L + 1. However there are many other ways of investigating the accuracy of your version of BASIC.

Index